The Physics of Organic Electronics

From molecules to crystals and polymers

The Physics of Organic Electronics

From molecules to crystals and polymers

Luís Alcácer

Instituto de Telecomunicações and Instituto Superior Técnico, Lisbon, Portugal

IOP Publishing, Bristol, UK

ISBN 978-0-7503-3347-4 (ebook)
ISBN 978-0-7503-3345-0 (print)
ISBN 978-0-7503-3348-1 (myPrint)
ISBN 978-0-7503-3346-7 (mobi)

DOI 10.1088/978-0-7503-3347-4

Version: 20220601

IOP ebooks

British Library Cataloguing-in-Publication Data: A catalogue record for this book is available from the British Library.

Published by IOP Publishing, wholly owned by The Institute of Physics, London

IOP Publishing, Temple Circus, Temple Way, Bristol, BS1 6HG, UK

US Office: IOP Publishing, Inc., 190 North Independence Mall West, Suite 601, Philadelphia, PA 19106, USA

In memory of my parents.

Contents

Preface

This book focuses on the physics behind the whole field of organic electronics. It addresses the fundamental physics of organic metals, superconductors, and semiconductors for organic electronics, as well as the physics of the most common organic electronic devices, in a comprehensive, global, and concise way.

The motivation for this book is to bring together the physics of organic conductors that preceded and led, in the 1980s, to organic electronic devices, and the physics that supports organic electronic technology. Books on organic electronics usually focus on the physics behind the technology and leave out the ideas and discoveries from the dawn of this field in the 1950s, revealing that organic conductors were the concretization of quasi one-dimensional condensed matter, opening the doors to exciting new science.

The book intends to provide a physics background of the highest value to any reader, graduate, senior undergraduate student, researcher, or engineer, who wishes to acquire a comprehensive overview of the foundations underlying organic electronics, an emerging new paradigm which promises sustainable technologies in such domains as lighting, solar cells, and especially in applications in the life sciences and healthcare.

The different but complementary aspects of this technology will be addressed in a logical sequence, from simple basic concepts to the level commonly appearing in the literature on the subject. The reader is expected to have a minimal background in soft and hard condensed-matter physics and elementary quantum mechanics, chemical physics, or materials science.

After a brief introduction to the field, an overview of the relevant electrical properties of materials is given, followed by short reviews of intermolecular interactions, structure, and scattering. This is followed by an overview of the main theories that deal with interacting electrons in quasi one-dimensional systems, including some rudiments of the elegant Luttinger liquid model, followed by two case studies, $(TMTSF)_2PF_6$ and the two-chain perylene–metal dithiolate series. These case studies illustrate how the theoretical models apply to organic conductors that exhibit most of the transport properties and mechanisms, phase transitions, and magnetic effects known in condensed matter as well as some new and surprising ones. The remaining chapters cover the physics more directly related to the technology: the physics of organic semiconductors for organic electronics, the physics of the most common devices (organic light emitting diodes, organic solar cells, and organic transistors) and molecular-scale electronic devices. Appendices on second quantization, phase transitions, and scaling laws are also included.

Luís Alcácer
February 2022

Acknowledgements

My thanks are first due to IOP Publishing for the invitation to write this book.

Warm thanks to Denis Jérome, Richard Friend, Jean-Paul Pouget, and Claude Bourbonnais for their long-time friendship and collaboration. Thanks are also due to George Malliaras, Robert Metzger, and the anonymous reviewers of the proposal for their encouraging words of support with which to face the challenge of covering the physics of the whole field of organic conductors and electronics.

I wish to express my profound gratitude to my co-workers of the Organic Electronics Group at the Instituto de Telecomunicações. Special thanks must go to Jorge Morgado for reviewing the entire final draft, to Ana Charas and Manuel Almeida for reading the draft at various stages and for their suggestions, and to Quirina Ferreira for her help on molecular-scale devices.

Finally, my thanks to Caroline Mitchell and Calum Sims at IOP Publishing for their assistance in the preparation of the manuscript in its final form, and to the Instituto de Telecomunicações for institutional support (FCT/MCTES through national funds and when applicable co-funded EU funds under the project UIDB/50008/2020).

Author biography

Luís Alcácer

Luís Alcácer is Emeritus Professor at Instituto Superior Técnico (IST-ULisboa) and senior researcher at Instituto de Telecomunicações. He received his PhD in Chemistry from the University of California, Riverside, in 1970. He was a researcher at the Nuclear Physics and Engineering Laboratory in Portugal and in 1975 became Full Professor at IST-ULisboa, where he lectured in the areas of Chemical Physics, Solid State Physics, and Computational Quantum Chemistry. His research interests have been focused on organic electronics. He is the author of *Introduction to Quantum Mechanics with Applications to Modern Computational Chemistry* (Lisbon: IST Press) 2007, 2014, 2019 (in Portuguese); *Electronic Structure of Organic Semiconductors* (San Rafael, CA: Morgan & Claypool) 2018; and co-author of *The Essence of Quantum Mechanics* (Lisbon: IST Press) 2020. He also authored *The Devil in the Quantum World*, (Lisbon: Gradiva) 2013 (in Portuguese), a popular science book. His scientific profile may be seen at https://www.it.pt/Members/Index/359. He lives in Estoril, Portugal.

Fundamental physical constants

Constant	Symbol	Value	Units
Speed of light in vacuum	c	299 792 458 (exact)	m s^{-1}
Planck's constant	h	6.626 070 15 × 10^{-34} (exact)	J Hz^{-1}
Reduced Planck's constant	$\hbar = h/2\pi$	1.054 571 817...×10^{-34} (exact)	J s
Elementary charge	e	1.602 176 634 × 10^{-19} (exact)	C
Boltzmann constant	k	1.380 649 × 10^{-23} (exact)	J K^{-1}
Avogadro constant	N_A	6.022 140 76 × 10^{23} (exact)	mol^{-1}
Vacuum electric permittivity	ε_0	8.854 187 817 8128(13) × 10^{-12}	F m^{-1}
Vacuum magnetic permeability	μ_0	1.256 637 062 12(19) × 10^{-6}	N A^{-2}
Electron mass	m_e	9.109 383 701 5(28) × 10^{-31}	kg
Proton mass	m_p	1.672 621 923 69(51) × 10^{-27}	kg

Notes: (i) CODATA International recommended 2018 values. (ii) The figures inside the brackets correspond to the uncertainty in the standard deviation in the last digits. https://physics.nist.gov/cuu/Constants/index.html.

Chapter 1

Introduction

In a restricted sense, *organic electronics* is an emerging technology based on organic semiconductors, as opposed to conventional electronics, which is based on inorganic semiconductors, especially silicon. So far, the most successful consumer organic electronic devices are the organic light-emitting diode (OLED) displays for smartphones and TV sets. Currently, the major efforts in research and development are centred on displays and lighting, solar cells, transistors, and sensors. But the potential applications of organic electronics are enormous, particularly in the life sciences and healthcare, due to their softness and biocompatibility, qualities of great value to the field of bioelectronics and in biosensors and implantable devices.

Technology is the application and consequence of science and, in this case, one can find relevant ideas and discoveries far back in time. The concept of *synthetic metal*, for example, can be traced back to 1911, when McCoy and Moore [1] proposed the preparation of composite metallic substances from non-metallic elements. The possibility of superconductivity in organic compounds was first considered by Fritz London in 1937 [2], when he suggested that aromatic rings containing delocalized electrons, as in anthracene, might exhibit supercurrents that flowed around the rings under an external magnetic field.

Two events are generally considered to mark the birth of the organic electronics field: the synthesis of the molecular salt of perylene–bromine, with a metal-like conductivity (10^{-3} to $1\ \mathrm{S\,cm^{-1}}$) in 1954 by Akamatu, Inokuchi, and Matsunaga [3], and the seminal paper of William Little, ten years later, in 1964, proposing a model organic compound which should exhibit superconductivity at room temperature [4].

Little's model was based on pairs of charge carriers moving along a conducting chain, bound by electronic excitations of polarizable side groups. The idea was rooted in the isotope effect of the Bardeen, Cooper, and Schrieffer (BCS) theory [5]. Little's proposed mechanism was based on attractive electron–electron interactions (excitons) rather then on the electron–phonon interactions of the BCS theory. Little's paper and the reports on several organic conductors in the late 1960s and early 1970s, in particular, the report on

doi:10.1088/978-0-7503-3347-4ch1

the metallic conductor TTF-TCNQ (where TTF is tetrathiafulvalene, and TCNQ is tetracyanoquinodimethane), and the serendipitous discovery of high conductivity in heavily doped polyacetylene [6] truly led to an era of great science, both in the search for high-temperature organic superconductivity and in the exploration of conducting polymers, which led to the award of the Nobel Prize in Chemistry 2000 to Alan Heeger, Alan MacDiarmid, and Hideki Shirakawa. Organic superconductivity was achieved in 1979 as the result of the collaboration between Klaus Bechgaard and Denis Jérome. The first organic superconductor was di-(tetramethyltetraselenafulvalene)-hexafluorophosphate, $(TMTSF)_2PF_6$, under a hydrostatic pressure of 12 kbar, which had a transition temperature of 0.9 K [7]. These discoveries and developments led to a new field of multidisciplinary research involving many branches of chemistry and physics, with particular emphasis on condensed-matter physics.

Organic conductors, in the broad sense, include crystalline organic metals and superconductors and organic semiconductors. Organic semiconductors for organic electronics can be small-molecule semiconductors or conjugated polymers. When crystalline, all these materials exhibit high anisotropy in their conductivity and other properties, due to their structure. They are either formed of stacks of flat organic conjugated molecules held together by π–π intermolecular interactions, or conjugated polymer chains. In both cases, electrons and holes can preferentially travel along the π-systems, making these materials *one-* or *quasi one-dimensional conductors.*

The present trend of organic electronic technology is focused on the design and synthesis of novel organic semiconductors and the optimization and invention of devices and systems, with the support of many branches of chemistry and physics, in particular, synthetic chemistry and photophysics. The more fundamental science of crystalline organic metals and superconductors is also very much alive and thriving. It is mainly focused on the search for new organic conductors, such as π–d systems and single-component molecular conductors, and on the behaviour of matter under extreme conditions, namely at very low temperatures, high pressures, and high magnetic fields, aiming at an understanding of quantum criticality, i.e. the physics of quantum critical points—points in the phase diagrams of materials where continuous phase transitions take place at absolute zero.

References

[1] McCoy H N and Moore W C 1911 *J. Am. Chem. Soc.* **33** 273–92
[2] London F 1937 *J. Chem. Phys.* **5** 837–8
[3] Akamatu H, Inokuchi H and Matsunaga Y 1954 *Nature* **173** 168–9
[4] Little W A 1964 *Phys. Rev.* **134** A1416–24
[5] Bardeen J, Cooper L N and Schrieffer J R 1957 *Phys. Rev.* **108** 1175–204
[6] Shirakawa H, Louis E J, MacDiarmid A G, Chiang C K and Heeger A J 1977 *J. Chem. Soc.* **16** 578–80
[7] Jérome D, Mazaud A, Ribault M and Bechgaard K 1980 *J. Phys. Lett.* **41** L95–98

Chapter 2

Electric properties of matter

This chapter introduces a few fundamental concepts relevant to the transport and magnetic properties of matter. The discussion is focused on topics which are relevant to organic conductors and that may not be found in textbooks on organic electronics.

2.1 The electron

The electron is a particle and a wave. It was discovered as a particle in 1897 by J J Thomson and was shown to behave as a wave in 1927 by Davisson and Germer and independently by G Thomson (son of J J Thomson), who confirmed that, as predicted by Louis de Broglie, electrons can be diffracted by crystals. Its dual character was clearly seen, for example, in the two-slit experiment reported by Tonomura *et al* using a low-intensity electron beam in time-lapse photography, in which the electrons accumulated in a screen, forming an interference pattern[1].

The electron is considered to be a point charge with rest mass of $m_e = 9.11 \times 10^{-31}$ kg, an electric charge of $-e$ (where $e = 1.60 \times 10^{-19}$ C is the elementary charge), and a spin of 1/2, which makes it a fermion. It is a constituent of all conventional matter and is responsible for all its physical and chemical properties. Electrons interact with each other and can even be bound in Cooper pairs and behave as bosons in superconductors.

The structures of materials depend on the way the electrons are distributed to form the various chemical bonds. In covalent bonds, pairs of electrons are shared by different atoms; in ionic solids, electrons are localized in ions that stick together by electrostatic forces; in common metals, the lattice ions are fixed and some electrons are free to move among them. In molecular solids, the molecules are held together by van der Waals interactions, which are due to fluctuations of the electronic densities.

[1] https://commons.wikimedia.org/wiki/File:Double-slit_experiment_results_Tanamura_2.jpg

doi:10.1088/978-0-7503-3347-4ch2

There are also hydrogen bonds and π–π interactions. The latter are the source of electrical conductivity in organic conductors.

The explanation of the electrical properties of materials is based on several approaches, or theories, some of which consider electrons to be particles, while others consider the electrons to be waves, in which case, the particle character is usually present as well. This is the case for band theory, in which the electrons propagate as waves in a periodic potential. There are also models in which the electrical properties are described by the motion of *quasiparticles*, i.e. 'dressed' electrons and other emergent particles.

In a material, the motion of electrons under a moderate electric field is approximately described by Ohm's law (1827), according to which, the resistance of a wire is given by $R = V/I$, where V is the applied voltage and I the current. This can also be written as $\mathbf{j} = \sigma\mathbf{E}$, where $\mathbf{j}$ is the current density, σ the conductivity, and $\mathbf{E}$ the applied electric field.

A successful model that explains the electrical properties of metals dates from 1900, before the dawn of quantum mechanics, and was proposed by Paul Drude [1]. Quantum mechanics was introduced into solid-state physics by Arnold Sommerfeld in 1928 [2].

In these models the conduction electrons, considered to be independent, form an electron gas, called the Fermi gas in the Sommerfeld theory. In spite of the simplicity of these models, they established some of the most important concepts of solid-state physics, which are still used today.

The electron–electron interactions can successfully be taken into account in many cases by the concept of a Fermi liquid introduced by Lev Landau in 1957. The concept of the Fermi liquid fails in the case of electrons interacting in one dimension, in which case it has to be replaced by a more sophisticated approach known as the Luttinger liquid theory.

2.2 The electron gas—Drude's theory

Drude developed his theory of metals from the kinetic theory of gases. He considered a gas of electrons, whose negative charges were compensated for by immobile positive ions. In today's understanding of this model, we assume that in a metal, the atoms release their valence electrons to form stable closed-shell ions which keep their equilibrium positions, forming an ionic lattice. The 'free' electrons form an *electron gas*, which obeys the kinetic theory.

In Drude's model, the electron–electron and electron–ion interactions are neglected. As in the kinetic theory of ideal gases, the particles are in random motion, colliding with each other. We can then define a *mean free path*, ℓ, as the mean distance between successive collisions—with the ions, Drude thought—considering a collision to be an instantaneous change in velocity with a probability per unit time $1/\tau$, where τ is the *scattering*, or *relaxation time*. The average velocity, v_0, can be derived from the kinetic energy, $1/2\ mv_0^2 = 3/2\ k_B T$. This implies that, at room temperature, the electrons are moving at velocities of the order of 10^5 m s^{-1}. Since every electron emerges from a collision in a random direction, there is no contribution from v_0 to charge transport. When an electric field is applied, the equation of motion is

$$\frac{d\mathbf{p}}{dt} = -e\mathbf{E} - \frac{\mathbf{p}}{\tau} \tag{2.1}$$

where $\mathbf{p} = m\mathbf{v}$ is the linear momentum, $-e\mathbf{E}$ is the Lorentz force, and $-\mathbf{p}/\tau$ is the collisional term. At equilibrium, $d\mathbf{v}/dt = 0$, the electrons drift with a velocity $\mathbf{v}_d = -e\mathbf{E}\tau/m$, and the resulting current density is $\mathbf{j} = -ne\mathbf{v}_d = ne^2\tau\mathbf{E}/m$. Considering that $\mathbf{j} = \sigma\mathbf{E}$, we get

$$\sigma = \frac{ne^2\tau}{m}$$

for the conductivity. Since the velocity is proportional to the applied field, we can define a quantity called the *mobility*, μ, such that $\mathbf{v}_d = -\mu\mathbf{E}$. We can then write $\mu = e\tau/m$, and

$$\sigma = n\mu e$$

for the conductivity. It is important to remember that the units of mobility are $[\mu] = \mathrm{m^2\,V^{-1}\,s^{-1}}$.

Relaxation times for metals such as copper are of the order of 10^{-14} to 10^{-15} s at room temperature, since their resistivity is in the range of $\mu\Omega$ cm. This implies mean free paths of about 1 to 10 Å, which are of the order of interatomic distances. For very pure crystalline metals at low temperatures, mean free paths are of the order of a few millimetres, corresponding to about ten million times the distance between atoms. Such values are clearly incompatible with a collisional model that describes the electrons' motion in a metal.

An oscillating electric field $\mathbf{E}(t) = \mathrm{Re}[\mathbf{E}\exp(-i\omega t)]$ implies a solution of equation (2.1) of the form $\mathbf{p}(t) = \mathrm{Re}[\mathbf{p}\exp(-i\omega t)]$. Using (2.1) we get $-i\omega\mathbf{p} = -e\mathbf{E} - \mathbf{p}/\tau$, leading to

$$\mathbf{p} = -\frac{e\,\mathbf{E}}{1/\tau - i\omega}.$$

Since $\mathbf{j} = -ne\mathbf{v}_d = -ne\ \mathbf{p}/m$, the current density is $\mathbf{j}(t) = \mathrm{Re}[\mathbf{j}\exp(-i\omega t)]$ with

$$\mathbf{j} = -ne\frac{\mathbf{p}}{m} = \frac{ne^2}{m}\frac{\mathbf{E}}{1/\tau - i\omega} = \frac{ne^2\tau}{m}\frac{1}{1 - i\omega\tau}\mathbf{E}. \tag{2.2}$$

In equation (2.2) we can easily recognize the expression for the DC conductivity, σ_0, as the first term on the right-hand side, and we can write

$$\sigma(\omega) = \frac{\sigma_0}{1 - i\omega\tau},$$

from which we can take the real part as

$$\mathrm{Re}[\sigma(\omega)] = \frac{\sigma_0}{1 + \omega^2\tau^2}.$$

This is the frequency-dependent *optical conductivity* in Drude's model.

For good metals and for visible light, the relaxation time, τ, is long in comparison with the period of the light, so that we may omit the collisional term in equation (2.1) and write $m\ddot{\mathbf{x}} = -e\mathbf{E}$. If $\mathbf{x}$ and $\mathbf{E}$ have the same time dependence, $\exp(-i\omega t)$, then $-\omega^2 m\mathbf{x} = -e\mathbf{E}$, and $\mathbf{x} = e\mathbf{E}/m\omega^2$. Since the dipole moment is $-e\mathbf{x} = -e^2\mathbf{E}/m\omega^2$, the polarization (dipole moment per unit volume) is $\mathbf{P} = -ne\mathbf{x} = -ne^2\mathbf{E}/m\omega^2$. We know that $\mathbf{D} = \varepsilon\mathbf{E} = \varepsilon_0\mathbf{E} + \mathbf{P}$, therefore we can write $\mathbf{D} = \varepsilon\mathbf{E} = \varepsilon_0\mathbf{E} - ne^2\mathbf{E}/m\omega^2 = (\varepsilon_0 - ne^2/m\omega^2)\mathbf{E}$, where $\mathbf{D}$ is the displacement field, ε is the permittivity, and ε_0 is the vacuum permittivity. For the relative permittivity or dielectric constant, $\varepsilon_r = \varepsilon/\varepsilon_0$, we can then write

$$\varepsilon_r(\omega) = 1 - \frac{ne^2}{\varepsilon_0 m\omega^2} = 1 - \frac{\omega_p^2}{\omega^2} \tag{2.3}$$

where $\omega_p = \sqrt{n\, e^2/\varepsilon_0\, m}$ is called the *plasma frequency*.

A metal is transparent to electromagnetic radiation when ε_r is positive and real, i.e. at frequencies higher than ω_p. Equation (2.3) explains why conjugated polymers are practically transparent in visible light. In a conjugated polymer chain there are two electrons per double bond, which implies an electron density of the order of 2×10^{21} to 4×10^{21} electrons per cm^3. If we accept the validity of Drude's model and take $n = 2 \times 10^{21}\text{cm}^{-3}$, the plasma frequency will be $\omega_p = 2.5 \times 10^{15}$ rad s^{-1}, which corresponds to a wavelength $\lambda_p = 747$ nm, higher than the wavelength of visible light (400 to 700 nm), i.e. $\omega > \omega_p$.

2.3 The Fermi gas—Sommerfeld's theory

As soon as it was discovered that electrons also behave as waves, it became obvious that Schrödinger's wave mechanics should be used to explain the properties of metals. This was done by Sommerfeld in 1928, based on the concept of free electrons moving in the metal (as in Drude's theory), but satisfying the Pauli exclusion principle and the Fermi–Dirac statistics.

The simplest approach is to consider that, in the ground state ($T = 0$), the electrons in a piece of metal behave as free and independent particles in a box, and to solve the Schrödinger equation, ignoring the ions except to compensate for the negative charges of the electrons:

$$-\frac{\hbar^2}{2m}\nabla^2\psi(\mathbf{r}) = E\,\psi(\mathbf{r}).$$

The wave functions, which are plane waves, and the energy levels are given by the equations

$$\psi_{\mathbf{k}}(\mathbf{r}) = \frac{1}{\sqrt{V}}e^{i\,\mathbf{k}\cdot\mathbf{r}} \qquad E(\mathbf{k}) = \frac{\hbar^2 k^2}{2m}. \tag{2.4}$$

$\mathbf{k} = (k_x, k_y, k_z)$ is the *wave vector* and $V = L_x \times L_y \times L_z$ is the volume of the box. Since $\mathbf{k}$ has three components, the energy is the sum of three terms. Each k

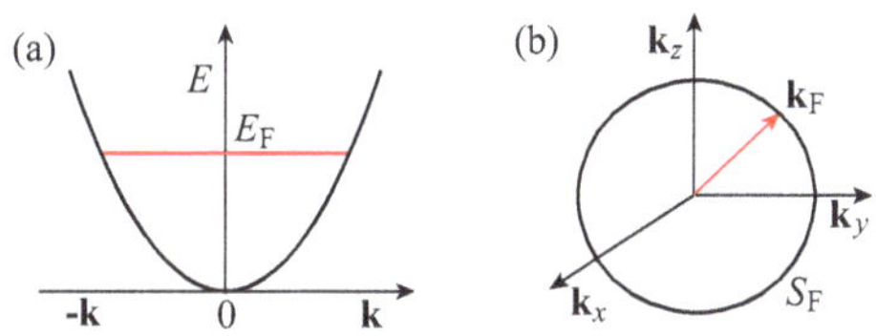

Figure 2.1. (a) Electron dispersion $E(\mathbf{k})$ (equation (2.4)) for the electron gas. E_F is the Fermi energy. (b) The Fermi surface S_F is a sphere of radius k_F.

component has to satisfy the Born–von Karman conditions $\psi(x + L_x) = \psi(x)$, $\psi(y + L_y) = \psi(y)$, and $\psi(z + L_z) = \psi(z)$, implying that $k_x = 2\pi\, n_x/L_x$, $k_y = 2\pi\, n_y/L_y$, and $k_z = 2\pi\, n_z/L_z$ (where n_x, n_y, and n_z are integers).

Although the energy levels are discrete (i.e. quantized) and are to be filled according to the *aufbau principle*, in a macroscopic piece of metal the levels are very close and the number of occupied levels is very large. We may then assume that they form a continuum and write the energy as a continuous function of $\mathbf{k}$, $E(\mathbf{k})$.

The wave vectors $\mathbf{k}$ define a k-space, also called *reciprocal space*, with dimensions of length^{-1}. Since $k_x = 2\pi\, n_x/L_x$, etc. each allowed state occupies a volume in reciprocal space $V_k = (2\pi)^3/V$, where V is the volume of the sample. We define a density of wave vector states, which is the number of states per unit volume in k-space, as

$$D(k) = \frac{\mathrm{d}N}{\mathrm{d}\mathbf{k}} = \frac{\mathrm{d}N}{\mathrm{d}^3k} = \frac{V}{(2\pi)^3} \qquad \mathrm{d}\mathbf{k} = \mathrm{d}^3k. \tag{2.5}$$

At $T = 0$, all levels are filled with two electrons each, up to the *Fermi level*, E_F, and the corresponding *Fermi wave vector*, k_F. Since, in real life, the number of occupied states is very large and each k-state occupies a very small volume, the total volume in k-space is essentially a sphere of radius $k_F = [2m\, E_F/(\hbar^2)]^{1/2}$ (figure 2.1). The surface in k-space that separates the occupied states from the empty states is called the *Fermi surface*, S_F.

2.3.1 Density of states

We define the density of states $D(E)$ as the number of states per unit of energy:

$$D(E) = \frac{\mathrm{d}N}{\mathrm{d}E} = \frac{\mathrm{d}N}{\mathrm{d}^3k}\frac{\mathrm{d}^3k}{\mathrm{d}E}.$$

The density of states can be calculated by considering the element of volume in k-space, d^3k, to be the product of the surface integral $\int_{S(E)} \mathrm{d}S$ over the constant energy surface $S(E)$ and the differential $\mathrm{d}k_\perp$ perpendicular to the surface, i.e. $\mathrm{d}^3k = \int_{S(E)} \mathrm{d}S \cdot \mathrm{d}k_\perp$. Using equation (2.5) for $\mathrm{d}N/\mathrm{d}^3k$, and introducing a factor of two to account for the two spin electron states, we get

$$D(E) = \frac{V}{4\pi^3} \int_{S(E)} \frac{\mathrm{d}S}{|\nabla_k E(\mathbf{k})|}.$$

The absolute value $|\nabla_k E(\mathbf{k})|$ is used to account for the fact that the density of states is always positive. This expression is general and can be used for any form of $E(\mathbf{k})$.

The density of states (DOS) is a very useful concept because it is experimentally accessible and can be the basis for the understanding of many electronic properties of solids, including amorphous materials and thin films.

2.3.2 Bosons and fermions

Quantum particles follow different distribution functions, depending on their spin quantum numbers. Particles with integer spin, such as photons ($S = 1$), follow the Bose–Einstein statistics, and are called *bosons*, while particles with semi-integer spin, such as electrons ($S = 1/2$), follow the Fermi–Dirac statistics, and are called *fermions*. These distribution functions are given by the following equations:

$$\text{Bose–Einstein: } f(E) = \frac{1}{\mathrm{e}^{(E-\mu)/k_\mathrm{B}T} - 1}; \qquad \text{Fermi–Dirac: } f(E) = \frac{1}{\mathrm{e}^{(E-\mu)/k_\mathrm{B}T} + 1}$$

where k_B is Boltzmann's constant, $f(E)$ is the average number of particles in a single-particle state with energy E, and μ is the chemical potential, which for fermions is defined as the Helmholtz free energy, F, of one particle added to a collection of N particles, i.e. $\mu = F_{N+1} - F_N$. Note that $\lim_{T\to 0}\mu = E_\mathrm{F}$, since the Fermi level is, by definition, the highest occupied energy level at $T = 0$. It can be shown that, even at room temperature, μ differs from E_F by only $\approx 0.01\%$. Therefore, for all practical purposes, we can assume that $\mu = E_\mathrm{F}$.

2.4 Electrons in a magnetic field

A magnetic field (see appendix C) applied perpendicularly to a current I_x generates a voltage V_H in the direction perpendicular to both the field and the current, as shown in figure 2.2. This was discovered by E H Hall in 1897 [3] and is called the Hall effect. To find out what happens, let us write the equation of motion for an electric charge $-e$ under an electric field $\mathbf{E} = (E_x, E_y, E_z)$ and a magnetic field $\mathbf{B} = (0, 0, B)$,

$$m\frac{\mathrm{d}\mathbf{v}}{\mathrm{d}t} = -e(\mathbf{E} + \mathbf{v} \times \mathbf{B}) - \frac{m\mathbf{v}}{\tau}. \tag{2.6}$$

In the stationary state, $\mathrm{d}\mathbf{v}/\mathrm{d}t = 0$. Multiplying both sides of (2.6) by τ/m and rearranging,

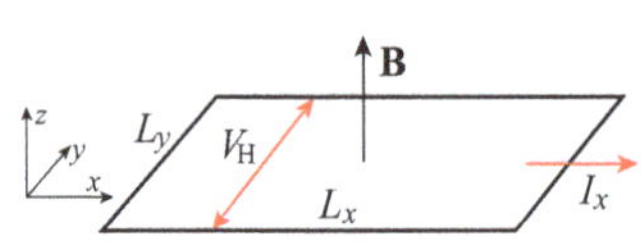

Figure 2.2. The classical Hall effect.

$$\mathbf{v} + \frac{e\tau}{m}\mathbf{v} \times \mathbf{B} = -\frac{e\tau}{m}\mathbf{E}. \tag{2.7}$$

We can now write equation (2.7) in matrix notation, multiply both sides of the equation by $-ne$, substitute $-ne\mathbf{v} = \mathbf{j}$, $\mathbf{v} = -e\tau\mathbf{E}/m$ and $ne^2\tau/m = \sigma_0$, and then substitute $eB/m = \omega_c$ and rearrange (see problem 2.2), to get

$$\begin{pmatrix} 1 & \omega_c\tau \\ -\omega_c\tau & 1 \end{pmatrix}\begin{pmatrix} j_x \\ j_y \end{pmatrix} = \sigma_0\begin{pmatrix} E_x \\ E_y \end{pmatrix} \tag{2.8}$$

where $\omega_c = eB/m$ is the *cyclotron frequency* and represents the frequency of the moving 'free' electron around the applied magnetic field.

From equation (2.8) we can write the two-dimensional resistivity tensor, $\boldsymbol{\rho}$:

$$\boldsymbol{\rho} = \frac{1}{\sigma_0}\begin{pmatrix} 1 & \omega_c\tau \\ -\omega_c\tau & 1 \end{pmatrix} = \begin{pmatrix} \rho_{xx} & \rho_{xy} \\ -\rho_{xy} & \rho_{yy} \end{pmatrix}.$$

For a current I_x flowing in the x direction, the Hall field which must compensate for the E_y component generated by the magnetic field will be $E_H = -E_y$. We define the Hall resistance as $R_{xy} = V_H/I_x = R_H B$, where R_H is the Hall coefficient:

$$R_H = \frac{E_H}{j_x B} = -\frac{E_y}{j_x B} = -\frac{\rho_{xy}}{B} = -\frac{\omega_c\tau}{B\sigma_0} = -\frac{1}{ne}. \tag{2.9}$$

Equation (2.9) and figure 2.2 show that if an electric current flows through a conductor in a magnetic field perpendicular to the direction of the current, the magnetic field exerts a transverse force on the moving electrons, pushing them to one side of the conductor. The build-up of charge produces a voltage across the two sides to balance the magnetic influence.

The sign of the Hall coefficient is the same as that of the charge carriers, $-e$. From the measurement of R_H, we get the charge carrier density, n, and its sign. It is $+e$ for 'holes'.

The longitudinal resistivity is $\rho_{xx} = 1/\sigma_0 = m/(ne^2\tau)$, and does not depend on the magnetic field—there is no magnetoresistance—in the framework of Drude's classical model.

2.4.1 Quantum-mechanical approach—Landau levels

The effect of a magnetic field on the motion of electrons in the xy plane can be described by the single-particle Schrödinger equation (leaving out the spin for the moment):

$$\left[\frac{1}{2m}(-i\hbar\nabla + e\mathbf{A})^2\right]\psi(x, y) = E\,\psi(x, y), \tag{2.10}$$

where $\mathbf{A}$ is the vector potential and m is the electron rest mass. If the magnetic field is in the z direction, $\mathbf{B} = (0, 0, B)$, which can be represented in the Landau gauge by $\mathbf{A} = (0, B\,x, 0)$. Equation (2.10) differs from the zero-magnetic-field equation because it has the vector potential $e\mathbf{A}$, where x appears explicitly. If we assume that the y component does not differ from the zero-field equation, the momentum operator in the y direction can be replaced by its eigenvalue $\hbar k_y$, and the solution of the Schrödinger equation can be written as

$$\psi(x, y) = e^{ik_y y}\phi(x).$$

If we define $x_0 = -\hbar k_y/eB = -k_y l_B^2$, where $l_B = \sqrt{\hbar/eB}$ is called the magnetic length, and introduce $\omega_c = eB/m$, we can write the equation for $\phi(x)$:

$$\left[-\frac{\hbar^2}{2m}\nabla_x^2 + \frac{1}{2}\omega_c^2 m(x - x_0)^2\right]\phi(x) = E\,\phi(x). \tag{2.11}$$

This equation is identical to the Schrödinger equation for the one-dimensional harmonic oscillator centred at x_0. The energy eigenvalues for equation (2.11) are, therefore,

$$E_\nu = \left(\nu + \frac{1}{2}\right)\hbar\omega_c \qquad \nu = 0, 1, 2, \ldots \tag{2.12}$$

The eigenfunctions are the product of the plane wave and the one-dimensional harmonic oscillator eigenfunctions:

$$\psi_{n,k_y}(x, y) \sim e^{ik_y y}H_n(x + k_y l_B^2)e^{(-x+k_y l_B^2)^2/2l_B^2},$$

where H_n are the Hermite polynomials. Part of this product is a Gaussian which localizes the wave function at $x_0 = -k_y l_B^2$ along the x direction, while it extends along the y direction.

The energy of an electron in a magnetic field, ignoring the spin component, is the energy of the oscillatory motion in the plane orthogonal to the field, which is quantized. The quantum of energy is $\hbar\omega_c$, ω_c being the *cyclotron frequency; x_0* is the *centre coordinate* and defines the centre of the cyclotron orbit. The integer ν defines the Landau levels.

Note that this is only the energy from the orbital motion of the electron. The spin contribution can then be added. Its value is $\pm(g/2)\mu_B B$, where g is the Landé g-factor and μ_B is the Bohr magneton, $\mu_B = e\hbar/2m$. We will ignore its energy contribution in the following explanation. Experimental values of the cyclotron frequency can be obtained from the absorption of electromagnetic waves, and g values can be measured by electron spin resonance spectroscopy. From these values, the effective mass m^* and g^* can be determined.

The situation we have described is that of electrons moving around the magnetic field in the z direction in a two-dimensional box of area $L_x L_y$. In a real sample with thickness t, the Hall resistance is $R_{xy} = \rho_{xy}L_y/L_y t = \rho_{xy}/t = -B/ent$, where we can combine the three-dimensional electron density and the thickness to get a two-dimensional electron density, $n\,t = n_s$. A two-dimensional box can be realized at a semiconductor/insulator interface, as in a GaAs/AlGaAs heterostructure, or in a MOSFET (metal–oxide–semiconductor field-effect transistor), where electrons are confined within a very thin layer of a nanometre or less.

Note that, since $x_0 = -\hbar k_y/eB$, $x - x_0 = x + \hbar k_y/eB$. This means that x_0 has to be within the 2D box, i.e. $0 < x_0 < L_x$ and, making $k_y = 2\pi\, n_y/L_y$ (see section 2.3), we can write

$$0 < n_y < L_x L_y \frac{e}{h} B = n_{y_{\max}}.$$

This defines the degeneracy of each Landau level, which is the maximum number of states n_y that can exist in the area $L_x L_y$ for each Landau level. Since the magnetic flux is the product of the magnetic field and the area perpendicular to it, $\Phi = BL_x L_y$, we can write

$$L_x L_y \frac{e}{h} B = \frac{\Phi}{\Phi_0} \qquad \text{with} \quad \Phi_0 = \frac{h}{e}.$$

$\Phi_0 = h/e$ is the (Dirac) *quantum of magnetic flux*. Note that CODATA defines the quantum of magnetic flux as $\Phi_0 = h/2e$, which comes from the Josephson effect in superconductors, where the minimum charge is $2e$, the charge of a Cooper pair. The integer $n_{y_{\max}}$ is the number of flux quanta in the sample. The degeneracy factor per unit area is, therefore,

$$N_L = \frac{n_{y_{\max}}}{L_x L_y} = \frac{eB}{h}. \tag{2.13}$$

This is the number of states per unit area in each Landau level. It means that each state occupies the area h/eB in real space corresponding to the area of a flux quantum.

2.4.2 Quantum Hall effects

If the two-dimensional electron density n_s (number of electrons per unit area) and the magnetic field B are adjusted to completely fill i Landau levels, then n_s has to satisfy the condition

$$n_s = i\, N_L. \tag{2.14}$$

From the classical Hall effect (equation 2.9), recall that the Hall coefficient $R_H = -1/ne$, where n is the 3D electron density. Since we are now dealing with a 2D system we will use n_s instead of n, and also equation (2.13) to get R_H:

$$R_H = -\frac{1}{n_s e} = -\frac{1}{iN_L e} = -\frac{h}{e^2 B}\frac{1}{i}$$

from which we can derive the transverse resistivity ρ_{xy} which is $\rho_{xy} = -R_H B$ (see equation 2.9):

$$\rho_{xy} = \frac{h}{e^2}\frac{1}{i} \qquad i = 1, 2, 3, \ldots \tag{2.15}$$

which means that the transverse or quantum Hall resistivity is quantized at critical values of the magnetic field. This was discovered by Klaus von Klitzing in

February, 1980. His measurements of the Hall resistance agreed with the calculated values of h/e^2i with high accuracy, and were independent of the geometry of the sample and insensitive to the microscopic details of the material. He realized that this resistance would be important for metrology and submitted a letter to Phys. Rev. Letters with the title 'Realization of a Resistance Standard based on Fundamental Constants', but it was not accepted on the grounds that there was no need for a more accurate resistor. He changed the title to 'New Method for High-Accuracy Determination of the Fine-Structure Constant Based on Quantized Hall Resistance' and it was finally published in August [4]. The resistance quantum is $R_K = h/e^2$, the von Klitzing constant. Its value is *exactly* $R_K = 25\,812.807\,45\ \Omega$. It is the practical standard for electrical resistance. Klaus von Klitzing was awarded the Nobel Prize in Physics in 1985. A 2017 review by von Klitzing [5] gives an interesting account of his discovery. Equation (2.15) is the expression for the *integer quantum Hall effect* as a function of i which is called the *filling factor* and is an integer if i energy levels of equation (2.12) are fully occupied. We can rewrite (2.14) to explicitly give the condition that i has to satisfy:

$$i = \frac{h\, n_s}{eB} \qquad i = 1, 2, 3, \ldots \tag{2.16}$$

Note that $i = 0$ corresponds to a state in which all Landau levels are empty. When condition (2.16) is satisfied, the conductivity σ_{xx} in the direction of the electric field becomes zero, since the electrons are moving perpendicularly to the electric field E_x and no diffusion originating from scattering in the x direction is possible. The theory of quantum transport in a 2D electron system under magnetic fields [6] predicts the broadening of the Landau levels and a σ_{xx} that is proportional to the square of the density of states at the Fermi level, implying $\sigma_{xx} = 0$ in the quantum Hall regime and quantized plateaux in the Hall resistivity ρ_{xy} (see figure 2.3). For the integral quantum Hall effect, each plateau is centred at $B = hn_s/ie = (n_s\, \Phi_0/i)$ and extends over a range of magnetic fields. The origin of the plateaux is disorder and their extension increases with the amount of disorder (within reason). While the Hall resistivity ρ_{xy} remains on a plateau, the longitudinal resistivity vanishes ($\rho_{xx} = 0$) and it spikes when ρ_{xy} jumps to the next plateau. These are the Shubnikov–de Haas oscillations caused by the jumps of the Fermi level across the gap between Landau levels. On a plateau, the Hall sample is a perfect conductor with $\rho_{xx} = 0$.

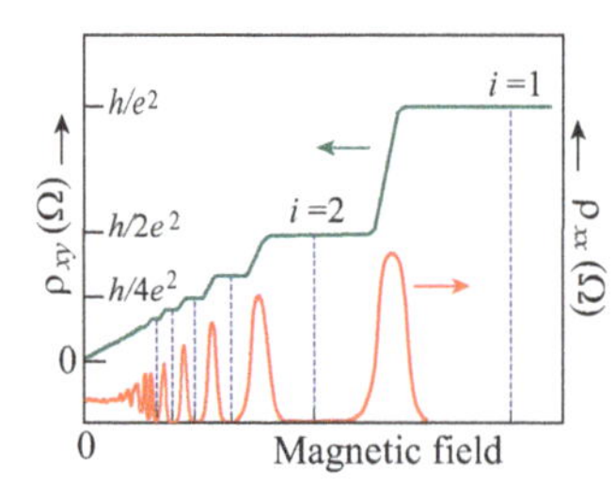

Figure 2.3. Typical features of the integer quantum Hall effect, ρ_{xy}, and magnetoresistivity, ρ_{xx}, as a function of the magnetic field.

However, due to the tensorial nature of the resistance in two dimensions, it is also a perfect insulator: $\sigma_{xx} = 0$. This can be seen from the relation between the resistivities and the conductivities, considering that $\mathbf{A}^{-1} = \mathbf{A}^{\dagger} / \det(\mathbf{A})$:

$$\sigma_{xx} = \frac{\rho_{xx}}{\rho_{xx}^2 + \rho_{xy}^2} \qquad \sigma_{xy} = \frac{-\rho_{xy}}{\rho_{xx}^2 + \rho_{xy}^2}.$$

Only when $\rho_{xy} = 0$ do we have the usual $\sigma_{xx} = 1/\rho_{xx}$.

The above approach to the integer quantum Hall effect is relatively naive, and it is amazing that it gives results with such accuracy that agree with simple fundamental equations. The existence of the quantum Hall plateaux and the behaviour of the longitudinal magnetoresistance is difficult to understand, although it is known to be related to the effect of disorder on the electronic wave functions, which leads to electron localization.

In 1982, Tsui, Störmer, and Gossard found to their great surprise an additional quantized Hall plateau of ρ_{xy} accompanied by a minimum in ρ_{xx} at 1/3 filling of the first Landau level [7]. They later found more new steps above and between the integers. All these plateaux can be expressed with the same equation (2.15), $\rho_{xy} = h/ie^2$, but in this case, i can be a fraction. They had discovered the *fractional quantum Hall effect*, but had no explanation for it. It was Robert Laughlin who first gave a phenomenological explanation, but it only applies to fillings where $i = 1/m$ and m is an odd integer. The 1998 Nobel Prize in Physics was awarded to Robert Laughlin, Horst Störmer, and Daniel Tsui 'for their discovery of a new form of quantum fluid with fractionally charged excitations'.

2.5 Beyond the electron gas model

In the Drude and Sommerfeld models the electrons were considered independent, ignoring the interactions between one electron and another. This is, of course, unrealistic. The approach most commonly used to overcome that problem is the Hartree–Fock or mean field approximation, which consists of considering the interactions of each electron with the other $N - 1$ reduced to a mean field potential which depends only on its coordinates. This is the basis of powerful computational methods that predict the electronic structure on scales from single molecules to complex solids. In crystals, this approximation is combined with Bloch's idea that the moving electrons can be described as waves with certain wavelengths, or momenta, travelling through the periodic potential generated by the crystal lattice [8] and which gave rise to *band theory*, the most common approach to describing the electronic structures of metals and semiconductors. The electronic structure of organic conductors and superconductors is primarily described in the context of an elementary tight binding approximation based on linear combinations of orbitals. Computational methods based on density functional theory are commonly used to describe the electronic structure of organic semiconductors in greater detail. However, the explanation of the properties of low-dimensional conductors requires sophisticated models such as the Luttinger liquid theory.

2.5.1 The Fermi liquid theory

In spite of the strength of the electron–electron interactions, the success of the independent electron approximation may be ascribed to the Fermi liquid theory developed by Lev Landau in 1957 [9, 10]. He showed that the thermodynamic properties of a Fermi gas and a Fermi liquid are very similar. The Fermi liquid is similar to a Fermi gas of collective excitations or *quasiparticles*, each with a different effective mass. A good example of the success of this theory is the specific heat, C_V, which has the same mathematical expression but with m replaced by m^* and provides a direct determination of m^*.

In the free electron gas, at $T = 0$, all states up to the Fermi level are occupied. Therefore, the occupation number n_k has a discontinuity at k_F and the amplitude of the discontinuity Z is one. The excitations of the system consist of the addition of particles with a well-defined energy $\varepsilon(k)$. The properties of the Fermi liquid are essentially similar, but the electrons are 'dressed' by their interactions with the other particles (electron–hole excitations). These entities are the *quasiparticles* which are considered as essentially free. Some properties, such as susceptibilities, are sensitive to residual interactions which can be described by the Landau parameters. The occupation number n_k still has a discontinuity at k_F (figure 2.4(a)), but its amplitude is less than one—the more interacting the system, the smaller the discontinuity. These quasiparticles have well-defined frequencies and momenta, $\omega = E(k)/\hbar$, and can be described by wave functions with a time dependence, $\exp(\mathrm{i}\,\omega\, t)$, but $E(k)$ is not the energy of a free electron. Since the energies of interest are close to the Fermi energy, we can linearize the dispersion in powers of $k - k_F$ and take only the first term for simplicity:

$$E(k) \simeq E(k_F) + \frac{\hbar^2 k_F}{m^*}(k - k_F)$$

where m^* is an effective mass. These quasiparticles have a finite lifetime shown by Landau to diverge in three-dimensional systems according to $1/E(k)^2$, which implies, for example, that the resistivity should be proportional to T^2.

A feature of the Fermi liquid theory is that it is very sensitive to the dimensionality of the electron system. In fact, in the 1950s, Sin-Itiro Tomonaga showed that the Fermi liquid theory would break down in one-dimensional systems. Later, in the 1960s, Joaquin Luttinger proposed a model of a one-dimensional metal in which the only stable low-lying excitations were quantized spin and charge waves that obeyed

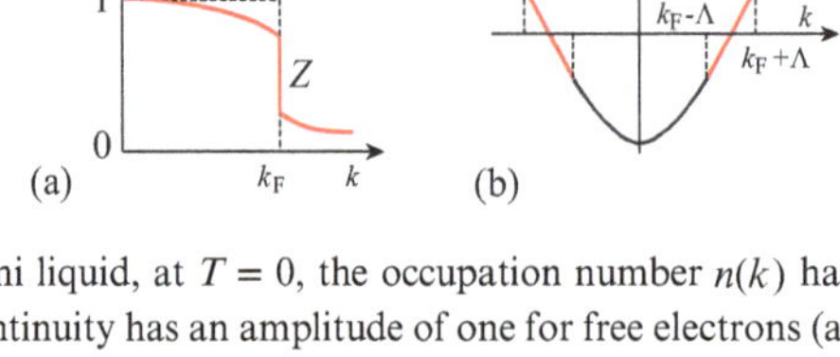

Figure 2.4. (a) For the Fermi liquid, at $T = 0$, the occupation number $n(k)$ has a discontinuity of amplitude $Z < 1$ at k_F, while the discontinuity has an amplitude of one for free electrons (a Fermi gas). (b) The linearized region of the energy dispersion in the vicinity of $\pm k_F$ (see text).

boson statistics. In addition, several properties of one-dimensional metals, such as their responses to electric and magnetic fields, are characterized by power-law singularities of the form $T^{-\alpha}$, where T is the temperature and α depends on the elastic constants of the system. The spin–charge separation and the power-law behaviour prevalent in one-dimensional electron systems are the key features of so-called Luttinger liquids. This topic will be addressed in more detail in chapter 5.

Further reading

Ashcroft N W and Mermin N D 1976 *Solid State Physics* (Philadelphia, PA: Saunders)

Douçot B, Pasquier V, Duplantier B, and Rivasseau V (ed) 2005 *The Quantum Hall Effect. Poincaré Seminar* 2004 (Basel: Birkhäuser)

Ezawa Z F 2013 *Quantum Hall Effects: Recent Theoretical And Experimental Developments* 3rd edn (Singapore: World Scientific)

Problems

2.1 A sample of PEDOT:PSS treated with 15 vol% DMSO has the following properties: conductivity: 570 S cm^{-1}, carrier mobility: 78 cm^2 V^{-1} s^{-1}, and a plasma frequency of 67 THz. Assume that the Drude model is approximately valid.

(a) Find out whether this polymer is transparent in the visible range.

(b) Calculate the charge density.

(c) Calculate the effective mass.

2.2 Derive equation (2.8) from (2.7).

References

[1] Drude P 1900 *Ann. Phys.* **306** 566–613
[2] Sommerfeld A 1928 *Z. Phys.* **47** 1–32
[3] Hall E H 1879 *Am. J. Math.* **2** 287–92
[4] Klitzing K v, Dorda G and Pepper M 1980 *Phys. Rev. Lett.* **45** 494–7
[5] Klitzing K 2017 *Annu. Rev. Condens. Matter Phys.* **8** 13–30
[6] Ando T 1974 *J. Phys. Soc. JapanJ. Phys. Soc. Japan* **37** 1233–7
[7] Tsui D C, Stormer H L and Gossard A C 1982 *Phys. Rev. Lett.* **48** 1559–62
[8] Bloch F 1928 *Z. Physik* **52** 555
[9] Landau L D 1957 *Sov. Phys. JETP* **3** 920–5
[10] Landau L D 1959 *Sov. Phys. JETP* **8** 361–2

Chapter 3

Intermolecular interactions

Organic conductors are usually molecular crystals or polymers, whose structures depend on intermolecular interactions, while the structures of inorganic metals and semiconductors depend on metallic or covalent bonds. This chapter is about those intermolecular interactions.

3.1 Electrostatic interactions

Electrostatic interactions play a crucial role in organic electronics, and they can usually be treated classically. One should recall that the chemical bond emerges from the Schrödinger equation by introducing the electrostatic potential; the quantum effects come from correlations, exchange interactions, and Pauli's exclusion principle. Computational methods based on Hartree–Fock and density-functional theory, as well as on Monte Carlo and molecular dynamics, can treat covalent and non-covalent interactions. However, here we will use a classical approach to get the physical picture of intermolecular interactions.

We could start by writing down the force between two charges and then find out the equations for the field, the electrical potential, the energy of interaction between the two charges, and so forth; however, to be more general, we will use Maxwell's equations. Electrostatics can be condensed into Maxwell's two equations containing the electric field, **E**, and in which the time derivatives are zero, i.e. the static case. In a given material, the macroscopic equations, also known as Maxwell's equations in matter, are as follows (in SI units):

$$\begin{aligned} \nabla \cdot \mathbf{D} &= \rho \qquad \text{Gauss's law} \\ \nabla \times \mathbf{E} &= 0 \qquad \text{Faraday's Law} \end{aligned} \tag{3.1}$$

where $\nabla = \frac{\partial}{\partial x}\mathbf{e}_x + \frac{\partial}{\partial y}\mathbf{e}_y + \frac{\partial}{\partial z}\mathbf{e}_z$ is a vector operator, which, when operating on a vector field, is the divergence of the field, giving a measure of the quantity of flux (or field lines) emanating from any point of the field. **D** is the *electric displacement*,

doi:10.1088/978-0-7503-3347-4ch3

$\mathbf{D} = \varepsilon\mathbf{E} = \varepsilon_0\mathbf{E} + \mathbf{P}$, where $\mathbf{E}$ is the electric field and $\mathbf{P}$ is the *polarization* (dipole moment per unit volume); ε is the permittivity, $\varepsilon = \varepsilon_r\varepsilon_0$, where ε_r is the relative permittivity (or dielectric constant) and ε_0 ($\varepsilon_0 = 8.854 \times 10^{-12}$ C^2 J^{-1}) is the vacuum permittivity; and ρ is the charge density.

From Gauss's law, one can derive Coulomb's law using Gauss's theorem

$$\int_V (\nabla \cdot \mathbf{E})\, \mathrm{d}\mathbf{r} = \oint_S \mathbf{E} \cdot \mathrm{d}\mathbf{A} = \frac{\rho}{\varepsilon} \tag{3.2}$$

meaning that the total electric flux from a closed surface S defining a unit volume V is equal to the charge ρ enclosed, divided by the permittivity $\varepsilon = \varepsilon_r\varepsilon_0$; $\mathrm{d}\mathbf{r}$ and $\mathrm{d}\mathbf{A}$ are the volume and area elements ($\mathrm{d}\mathbf{A} = \mathbf{n}\,\mathrm{d}S$), respectively. To make things easier, one can just consider a single point charge q in a spherical unit volume, and write

$$\oint_S \mathbf{E} \cdot \mathrm{d}\mathbf{A} = \mathbf{E} \oint_S \mathrm{d}A = \mathbf{E}\, 4\pi r^2 = \frac{q}{\varepsilon},$$

from which one can get the field

$$\mathbf{E} = \frac{1}{4\pi\varepsilon}\frac{q}{r^2}\,\mathbf{e}_r \tag{3.3}$$

a distance r away from the charge and directed along the vector $\mathbf{r}$ emanating radially from the charge; $\mathbf{e}_r$ is the unit vector along $\mathbf{r}$.

Since the electric field $\mathbf{E}$ at a given point $\mathbf{r}$ is the force, $\mathbf{F} = q\,\mathbf{E}$, acting on a unit charge at that point, we can write Coulomb's law, which gives the electrostatic force between two charges, q_1 and q_2, at a distance r in a material with relative permittivity of ε_r, as follows:

$$\mathbf{F} = \frac{1}{4\pi\varepsilon_r\varepsilon_0}\frac{q_1 q_2}{r^2}\mathbf{e}_r. \tag{3.4}$$

From Faraday's law and Stokes' theorem we can get

$$\oint_\Gamma \mathbf{E} \cdot \mathrm{d}\mathbf{s} = \int_S (\nabla \times \mathbf{E}) \cdot \mathrm{d}\mathbf{A}$$

which states that the circulation of a field around a loop Γ is equal to the surface integral of the normal component of $\nabla \times \mathbf{E}$. So, if $\nabla \times \mathbf{E}$ is zero, the circulation around any loop is zero. This implies that if we choose any two points on a closed curve, the line integral depends only on the position of the two points, not on the path taken. This leads to the idea of a *potential* ϕ. The field $\mathbf{E}$ is the gradient of this potential, i.e.

$$\mathbf{E} = -\nabla\phi.$$

The potential difference between two points a and b is therefore

$$\phi_b - \phi_a = -\int_a^b \mathbf{E} \cdot \mathrm{d}\mathbf{s}. \tag{3.5}$$

The SI units are: $[\mathbf{E}] = \mathrm{V\,m^{-1}}$ and $[\phi] = \mathrm{V}$. The minus sign means that a positive charge will move from a higher potential to a lower potential in an electric field.

The energy of a charge q in an electric potential ϕ is

$$W = q\,\phi.$$

The energy of interaction between the two charges is defined as the work required to bring the two charges from an infinite separation to a distance r from each other:

$$V(r) = -\int_{\infty}^{r} \mathbf{F} \cdot \mathrm{d}\mathbf{s} = \frac{1}{4\pi\varepsilon_r\varepsilon_0}\frac{q_1\,q_2}{r}.$$

Two separate charges $+q$ and $-q$ generate a dipole moment $\boldsymbol{\mu} = q\,\mathbf{r}$, in which $\mathbf{r}$ is the distance vector from the position of the negative charge to the positive charge. The dipole moment has SI units of $[\mu] = \mathrm{C\,m}$ (coulomb × meter); chemists, however, frequently use the debye: $1\mathrm{D} = 3.336 \times 10^{-30}\mathrm{C\,m}$. In an electric field $\mathbf{E}$, the dipole moment aligns with the field. We should also consider that atoms and molecules contain electrons and nuclei, i.e. both negative and positive electric charges, and that an external electric field will distort the distribution of those electric charges, inducing a dipole moment. If the field is not too strong, the induced dipole will be proportional to the field, i.e. $\boldsymbol{\mu}_{\mathrm{ind}} = \alpha\mathbf{E}$, where α is the polarizability.

3.1.1 Dipolar and van der Waals interactions

Van der Waals attractive interactions exist between all kinds of atom and molecule. They are the sum of three terms: (i) the *Keesom* (orientation) interaction between permanent dipoles of polar molecules, (ii) the *Debye* (induction) interaction between the permanent dipoles of polar molecules and dipoles they induce in identical molecules and in non-polar molecules, and (iii) the *London* (dispersion, or induced dipole–induced dipole) interactions between all types of molecule.

Although the magnitude of these interactions varies according to the type of molecule, they all scale as $1/r^6$. They are short range, as compared to interactions between ions and ion–dipole interactions. We will now describe each type of interaction in some detail, starting with the derivation of the electric potential and field due to a set of charges.

3.1.1.1 Electric field of a dipole

We need to derive the equations for the electric potential and field at a given point arising from a distant set of charges (e.g. a molecule). Take, for example, a distribution of charges q_i in a small piece of matter with a dielectric constant (relative permittivity) of ε_r and a faraway point $P(\mathbf{r})$, as shown in figure 3.1. Imagine a point charge q at the origin. From equations (3.5) and (3.3), we can calculate the potential at $P(\mathbf{r})$ due to that charge, considering that the potential at infinity is zero:

$$\phi(\mathbf{r}) = -\int_{\infty}^{r} \mathbf{E} \cdot \mathrm{d}\mathbf{s} = -\frac{1}{4\pi\varepsilon_r\varepsilon_0}\int_{\infty}^{r} \frac{q}{r^2}\mathrm{d}r = \frac{1}{4\pi\varepsilon_r\varepsilon_0}\frac{q}{r}.$$

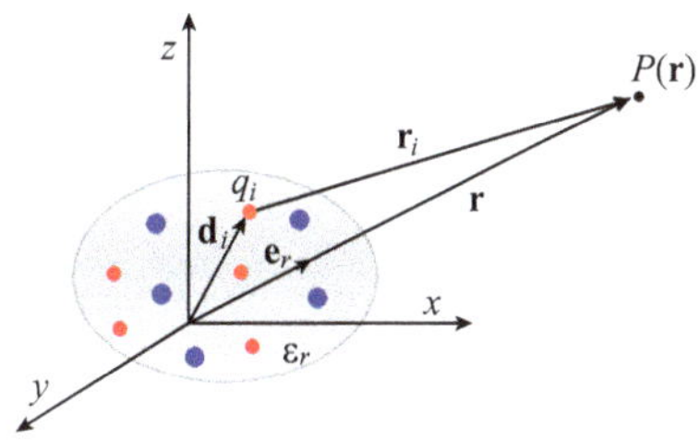

Figure 3.1. A set of charges q_i and a point $P(\mathbf{r})$ a large distance away.

Let each charge q_i in the set be at a distance $\mathbf{d}_i$ from the origin and suppose that the distance $\mathbf{r}$ of point $P(\mathbf{r})$ is much larger than any $\mathbf{d}_i$. The potential due to the set of charges is

$$\phi(\mathbf{r}) = \frac{1}{4\pi\varepsilon_r\varepsilon_0}\sum_i \frac{q_i}{r_i}. \tag{3.6}$$

If $\mathbf{e}_r$ is the unit vector in the direction of $\mathbf{r}$, then we can make the approximation

$$r_i \approx r - \mathbf{d}_i \cdot \mathbf{e}_r.$$

Since $d_i \ll r$, we can expand $1/r_i$ about $1/r$ in powers of d_i/r,

$$\frac{1}{r_i} \approx \frac{1}{r}\left(1 + \frac{\mathbf{d}_i \cdot \mathbf{e}_r}{r} + \dots\right)$$

where we leave out the higher-order terms. Substituting this into (3.6), we get

$$\phi(\mathbf{r}) = \frac{1}{4\pi\varepsilon_r\varepsilon_0}\left(\frac{\sum\limits_i q_i}{r} + \sum_i q_i\frac{\mathbf{d}_i \cdot \mathbf{e}_r}{r^2} + \dots\right). \tag{3.7}$$

Equation (3.7) is the multipole expansion of the potential due to a charge distribution, in which the first term is the contribution of the total charge distribution, the second term is the contribution of the dipole moment $\boldsymbol{\mu} = \sum_i q_i\, \mathbf{d}_i$, a third term would be the contribution of the quadrupole moment, and so forth. Another way would be to expand the potential in terms of the spherical harmonics, but the result would be the same. We can therefore write the dipole potential as

$$\phi(\mathbf{r}) = \frac{1}{4\pi\varepsilon_r\varepsilon_0}\frac{\boldsymbol{\mu} \cdot \mathbf{e}_r}{r^2} = \frac{1}{4\pi\varepsilon_r\varepsilon_0}\frac{\boldsymbol{\mu} \cdot \mathbf{r}}{r^3}.$$

We can now obtain the field, considering that $\mathbf{E} = -\nabla\phi$.

Let us write $\boldsymbol{\mu} \cdot \mathbf{r} = \sum_{i=x,y,z}\mu_i r_i = \mu_i r_i$ (using Einstein's convention that whenever an index appears twice in a product, it means a sum over all its values) and, therefore,

$$\phi(\mathbf{r}) = \frac{1}{4\pi\varepsilon_r\varepsilon_0}\frac{\mu_i r_i}{r^3}.$$

The components of the field can now be obtained using i and j for x, y, and z and considering $\nabla = \frac{\partial}{\partial x}\mathbf{e}_x + \frac{\partial}{\partial y}\mathbf{e}_y + \frac{\partial}{\partial z}\mathbf{e}_z$:

$$E_j = -\frac{\partial}{\partial r_j}\phi = -\frac{1}{4\pi\varepsilon_r\varepsilon_0}\mu_i\frac{\partial}{\partial r_j}\left(\frac{r_i}{r^3}\right). \tag{3.8}$$

To obtain the partial derivative, it is useful to decompose this into

$$\frac{\partial}{\partial r_j}\left(\frac{r_i}{r^3}\right) = \frac{\partial r_i}{\partial r_j}\frac{1}{r^3} + r_i\frac{\partial}{\partial r_j}\frac{1}{r^3}. \tag{3.9}$$

The first term gives $\partial r_i/\partial r_j = \delta_{ij}$, where δ_{ij} is the Kronecker symbol ($\delta_{ij} = 1$ if $i = j$; $\delta_{ij} = 0$ if $i \neq j$) or, $\partial x/\partial x = 1$, $\partial x/\partial y = 0$, etc. For the second term we have

$$\frac{\partial}{\partial r_j}\frac{1}{r^3} = \left(\nabla\frac{1}{r^3}\right)_j = -3\left(\frac{\mathbf{e}_r}{r^4}\right)_j = -3\frac{r_j}{r^5}.$$

Equation (3.9) becomes

$$\frac{\partial}{\partial r_j}\left(\frac{r_i}{r^3}\right) = \frac{1}{r^3}\left(\delta_{ij} - 3\frac{r_i r_j}{r^2}\right).$$

Finally, we can obtain the field components according to (3.8):

$$E_j = -\frac{1}{4\pi\varepsilon_r\varepsilon_0}\frac{1}{r^3}\left(\mu_i\,\delta_{ij} - 3\frac{(\mu_i\,r_i)r_j}{r^2}\right).$$

Using the Kronecker symbol combined with the sum over all i's, which selects the term with $i = j$, and considering that $\mu_i r_i = \boldsymbol{\mu}\cdot\mathbf{r}$, we get

$$E_j = \frac{1}{4\pi\varepsilon_r\varepsilon_0}\frac{1}{r^3}\left(3\frac{(\boldsymbol{\mu}\cdot\mathbf{r})r_j}{r^2} - \mu_j\right).$$

Putting all the components together, we get the field due to a dipole $\boldsymbol{\mu}$ at a distance $\mathbf{r}$:

$$\mathbf{E}(\mathbf{r}) = \frac{1}{4\pi\varepsilon_r\varepsilon_0}\frac{1}{r^3}\left[3\,\frac{(\boldsymbol{\mu}\cdot\mathbf{r})\mathbf{r}}{r^2} - \boldsymbol{\mu}\right]. \tag{3.10}$$

We can now use this expression for the field to get the energy of interaction between two dipoles (or two polar molecules), or between a dipole and an induced dipole (or a polar molecule and a non-polar molecule).

3.1.1.2 Interaction between two polar molecules

The main components of the van der Waals interaction are shown in figure 3.2. One of the components is the Keesom (or dipole–dipole) interaction between two polar molecules.

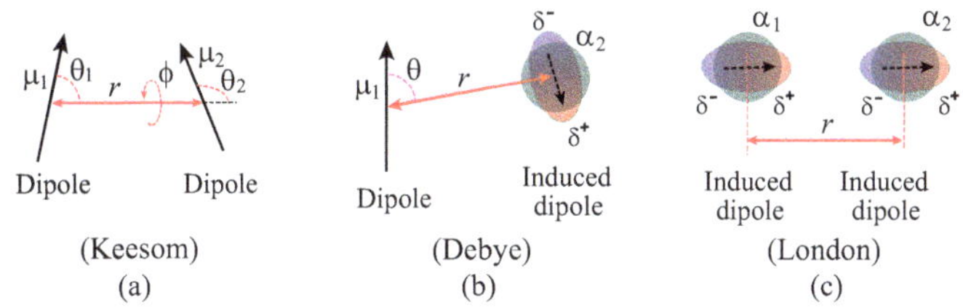

Figure 3.2. Intermolecular electrostatic interactions. (a) Dipole–dipole interaction (Keesom). (b) Dipole–induced dipole interaction (Debye). (c) Induced dipole–induced dipole (London).

To derive the energy of the dipole–dipole interaction, we may start from the equation for the potential energy of a dipole in an external field, $V_{dd} = -\boldsymbol{\mu} \cdot \mathbf{E}$ and use equation (3.10):

$$V_{\mathrm{dd}} = \frac{1}{4\pi\varepsilon_{\mathrm{r}}\varepsilon_0}\left[\frac{\boldsymbol{\mu}_1 \cdot \boldsymbol{\mu}_2}{r^3} - 3\frac{(\boldsymbol{\mu}_1 \cdot \mathbf{r})(\boldsymbol{\mu}_2 \cdot \mathbf{r})}{r^5}\right]. \tag{3.11}$$

This equation can also be written in terms of the angles between the dipole vectors and the vector connecting them, θ_1 and θ_2 (see figure 3.2(a)), and ϕ, which is the angle between the projections of the dipole moments on a plane perpendicular to the line through the centres:

$$V_{\mathrm{dd}}(r, \theta_1, \theta_2, \phi) = -\frac{1}{4\pi\varepsilon_{\mathrm{r}}\varepsilon_0}\frac{\mu_1\mu_2}{r^3}(2\cos\theta_1\cos\theta_2 - \sin\theta_1\sin\theta_2\cos\phi). \tag{3.12}$$

Equation (3.12) applies to the interaction between two polar molecules at a fixed orientation defined by the four parameters r, θ_1, θ_2, and ϕ. This may be the situation in a crystalline material. However, we are interested in the case in which the molecules can rotate relative to each other about the three angles, and we know that the energy of interaction depends on the relative orientation, being minimized when the dipoles are aligned, i.e. $\theta_1 = \theta_2 = 0$. We also know that any system will tend to reach a thermal equilibrium corresponding to the Boltzmann distribution, which is given by the thermal average over all angles:

$$V_{\mathrm{dd}}(r) \equiv \langle V\,\mathrm{e}^{-V/k_{\mathrm{B}}T}\rangle \equiv \frac{\int_0^{2\pi}\mathrm{d}\phi\int_{\theta_1=0}^{\pi}\int_{\theta_2=0}^{\pi} V\,\mathrm{e}^{-V/k_{\mathrm{B}}T}\sin\theta_1\,\mathrm{d}\theta_1\sin\theta_2\,\mathrm{d}\theta_2}{\int_0^{2\pi}\mathrm{d}\phi\int_{\theta_1=0}^{\pi}\int_{\theta_2=0}^{\pi}\mathrm{e}^{-V/k_{\mathrm{B}}T}\sin\theta_1\,\mathrm{d}\theta_1\sin\theta_2\,\mathrm{d}\theta_2}.$$

A calculation of the integrals leads to the final equation:

$$V_{\mathrm{dd}}(r) \approx -2\left(\frac{\mu_1\mu_2}{4\pi\varepsilon_{\mathrm{r}}\varepsilon_0}\right)^2\frac{1}{3\,k_{\mathrm{B}}T}\frac{1}{r^6}$$

which is the thermally averaged dipole–dipole, or Keesom interaction, one of the main contributors to the van der Waals interaction.

3.1.1.3 Interaction between a polar molecule and a non-polar molecule

The electric field of the dipole of a polar molecule induces a dipole in a non-polar molecule, aligned with the field, $\boldsymbol{\mu}_{\text{ind}} = \alpha\mathbf{E}$. This is called the Debye interaction. Using figure 3.2(b) and equation (3.10) we can find the E_r and E_θ components of the field of the polar molecule acting on the induced dipole of the non-polar molecule:

$$\mathbf{E}_{\mu_1}(r, \theta) = \frac{1}{4\pi\varepsilon_r\varepsilon_0}\frac{\mu_1}{r^3}(2\cos\theta\,\mathbf{e}_r + \sin\theta\,\mathbf{e}_\theta) \tag{3.13}$$

where $\mu_1 = |\boldsymbol{\mu}_1|$. The Debye interaction energy is then

$$V_{\text{D}}(r, \theta) = -\boldsymbol{\mu}_{\text{ind}} \cdot \mathbf{E}_{\mu_1} = -\alpha\, E_{\mu_1}^2. \tag{3.14}$$

From (3.13) we can get $E_{\mu_1}^2$:

$$E_{\mu_1}^2 = \left(\frac{1}{4\pi\varepsilon_r\varepsilon_0}\frac{\mu_1}{r^3}\right)^2(2\cos\theta\,\mathbf{e}_r + \sin\theta\,\mathbf{e}_\theta)^2 = \left(\frac{1}{4\pi\varepsilon_r\varepsilon_0}\frac{\mu_1}{r^3}\right)^2(3\cos^2\theta + 1).$$

Substituting this into (3.14), we get $V_{\text{D}}(r, \theta)$:

$$V_{\text{D}}(r, \theta) = -\alpha_2\left(\frac{1}{4\pi\varepsilon_r\varepsilon_0}\frac{\mu_1}{r^3}\right)^2(3\cos^2\theta + 1)$$

which should be angle averaged. It can be shown that $\langle\cos^2\theta\rangle = 1/3$, therefore we get

$$V_{\text{D}}(r) = \langle V_{\text{D}}(r, \theta)\rangle = -2\frac{\mu_1^2\alpha_2}{(4\pi\varepsilon_r\varepsilon_0)^2}\frac{1}{r^6}.$$

3.1.1.4 Interaction between two non-polar molecules

The interaction between two non-polar molecules, also called the London or dispersion interaction, is the result of fluctuations in the charge densities of the interacting atoms or molecules, and it can be described as an interaction between induced dipoles. The term 'dispersion' is used because this interaction is associated with the dispersion of light in materials. The attractive character of this interaction derives from the tendency of the polarizabilities to oscillate in phase along the line joining the atoms or molecules, as shown in figure 3.2(c). London derived the equation for the dispersion interaction energy of two neutral atoms as oscillating 1*s* orbitals with frequencies ν_1 and ν_2 [1]. London's equation gives the dispersion interaction in terms of the frequencies ν_1 and ν_2, but these frequencies multiplied by Planck's constant can be approximated to the first ionization potentials, i.e. $h\nu_1 = I_1$ and $h\nu_2 = I_2$. In this way we can obtain London's equation as it usually appears in the literature:

$$V_{\text{L}} = -\frac{3}{2}\frac{\alpha_1\alpha_2}{(4\pi\varepsilon_r\varepsilon_0)^2}\frac{I_1 I_2}{I_1 + I_2}\frac{1}{r^6}$$

where α_1 and α_2 are the polarizabilities of the two molecules.

3.1.2 The van der Waals interaction and the Lennard-Jones potential

The three contributions to the van der Waals interaction should then be added together:

$$V_{\mathrm{vdW}}(r) = V_{\mathrm{K}}(r) + V_{\mathrm{D}}(r) + V_{\mathrm{L}}(r)$$

noting that the London, or dispersion, interaction is always present (see table 3.1).

The repulsive interactions vary more rapidly than the attractive potentials. A common model for the total intermolecular interaction is the Lennard-Jones or '6–12' potential,

$$V_{\mathrm{LJ}}(r) = 4\, w_0 \left[\left(\frac{\sigma}{r}\right)^{12} - \left(\frac{\sigma}{r}\right)^{6} \right] \tag{3.15}$$

where w_0 is the depth of the potential well, which occurs at $r_0 = 2^{1/6}\sigma$, σ being the separation at which $V = 0$. This potential is shown in figure 3.3. So far, we have been considering the interaction between two molecules in a material with a relative permittivity of ε_r. In a molecular crystal the molecules are held together in equilibrium positions by the balance between attractive van der Waals interactions (and eventual hydrogen bonds) and the repulsive forces between atoms due to the interaction of their electron shells. The shortest possible distance between two atoms belonging to two different molecules is the sum of their van der Waals radii, an idea

Table 3.1. The main types of attractive intermolecular interaction which contribute to the van der Waals interaction, and the orders of magnitude of the interaction energies, V.

Interaction Type	Potential energy	V (kJ mol^{-1})
Dipole–dipole (Keesom)	$V_{\mathrm{K}} = -\frac{C_{\mathrm{K}}}{r^6};\ C_{\mathrm{K}} = \frac{2}{3}\frac{1}{k_{\mathrm{B}}T}\left(\frac{\mu_1\mu_2}{4\pi\varepsilon_r\varepsilon_0}\right)^2$	0.3
Dipole–ind. dipole (Debye)	$V_{\mathrm{D}} = -\frac{C_{\mathrm{D}}}{r^6},\ C_{\mathrm{D}} = 2\frac{\mu_1^2\alpha_2}{(4\pi\varepsilon_r\varepsilon_0)^2}$	0.8
Ind. dipole–ind. dipole (London)	$V_{\mathrm{L}} = -\frac{C_{\mathrm{L}}}{r^6};\ C_{\mathrm{L}} = \frac{3}{2}\frac{\alpha_1\alpha_2}{(4\pi\varepsilon_r\varepsilon_0)^2}\frac{I_1 I_2}{I_1 + I_2}$	2

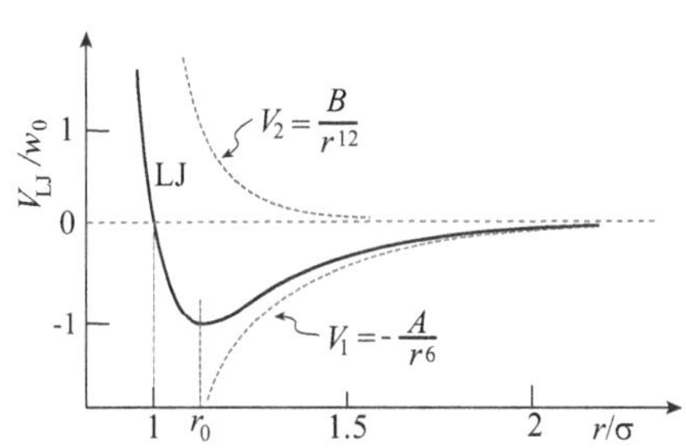

Figure 3.3. The Lennard-Jones (LJ) potential (equation (3.15)) showing the r^{-12} repulsive core potential and the r^{-6} attractive tail. The LJ potential passes through zero at $r/\sigma = 1.0$ and has a minimum at $r_0/\sigma = 2^{1/6} \approx 1.12$, where σ is the separation for which $V = 0$.

put forward by Linus Pauling in his book *The Nature of the Chemical Bond* [2]. From the analysis of the structures of a large number of molecular crystals it has been possible to collect the characteristic radius for each type of atom. Such values are presently available for most elements of the periodic table.

Simple formulae can be used to estimate the van der Waals intermolecular interaction energy in a crystal, based on the sums of the atom–atom and, eventually, the atom–bond and bond–bond contributions, such as the following [3]:

$$V_{\mathrm{vdW}} = \sum_{ij}\left[A_{ij}\exp(-\alpha_{ij}r_{ij}) - \frac{C_{ij}}{r_{ij}^{6}}\right] \tag{3.16}$$

where r_{ij} is the distance between atom i in one molecule and atom j in the other, and C_{ij}, A_{ij}, and α_{ij} are standard parameters independent of the atomic species i and j (see [4]).

3.2 Hydrogen bond

The recommended definition of 'hydrogen bond' given by the International Union of Pure and Applied Chemistry (IUPAC) is the following [5]: 'The hydrogen bond is an attractive interaction between a hydrogen atom from a molecule or a molecular fragment X–H in which X is more electronegative than H, and an atom or a group of atoms in the same or a different molecule, in which there is evidence of bond formation.'

This bond is usually represented by three dots, as in X–H···Y. The energy of hydrogen bonds in solids can vary between 0.2 and 40 kcal mol^{-1} and there are many different types, as described in a recent review [6]. Hydrogen bonds are ubiquitous in nature, they are innate in water and biology, and they are particularly important in organic electronics [7].

There are many models for the water molecule in all its phases as well as for the H-bond in water and in other molecular systems. Although the H-bond is now believed to be purely electrostatic, there is no equation that accurately predicts the interaction potential. In liquid water the energies of hydrogen bonds tend to follow a $1/r^2$ distance dependence, as expected for the attractive charge–dipole interaction, in which the charge and the dipole are almost aligned, i.e. near the $O^{\delta-}-H^{\delta+}\cdots O^{\delta-}$ configuration. An example of this is shown in figure 3.4, in which the geometry of the H-bond is based on data reported by Modig *et al* [8] as determined by ^{1}H NMR

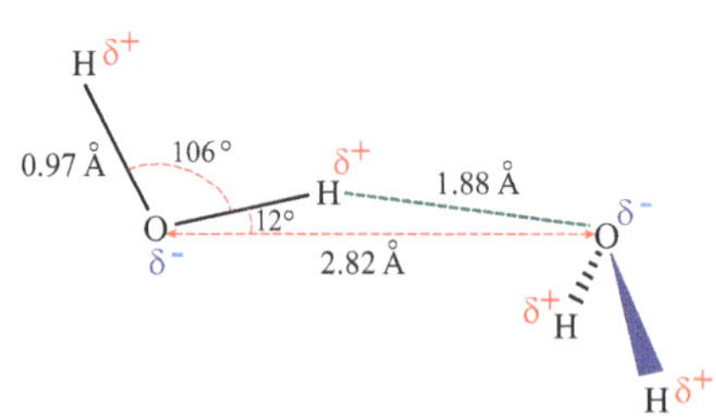

Figure 3.4. Model of the hydrogen bond between two water molecules in liquid water.

(nuclear magnetic resonance) line shape and density functional theory (DFT) calculations. A particular description of the H-bond given by Israelachvili in his book [9] proposes the following equation for the energy of the electrostatic attraction between a charge and a dipole:

$$V_{\mathrm{Hb}}(r) = -\frac{\delta^- \mu_{\mathrm{HO}} \cos\theta}{4\pi\varepsilon_r\varepsilon_0\, r^2}$$

where the charge is $\delta^- = -0.24\ e$, the distance between the charge and the centre of the dipole is 2.26 Å, the angle between the dipole and the H-bond is taken to be $\theta = 180°$, and $\varepsilon_r = 1$. The values for the charges and distances were taken from a model of the water molecule based on classical Monte Carlo statistical mechanics calculations [10]. The result of this simple calculation is $V_{\mathrm{Hb}}(r) \approx 3.4 \times 10^{-20}$ J = 4.9 kcal mole^{-1}, which agrees very well with the value of 4.5 kcal mole^{-1} for the water–water H-bond found in the literature. The partial atomic charges in solids are usually around +0.4 e on each H atom and −0.8 e on the O atom [6].

In crystals, hydrogen bonds have well-defined geometries and there are databases available, such as the *Cambridge Structural Database*, which provide a large amount of experimental data to help in the analysis of H-bond geometries.

3.3 Interactions involving π-electron systems

Organic electronic materials generally owe their electronic and optical properties to the presence of conjugated bonds, i.e. systems of connected p atomic orbitals which form molecular π orbitals with delocalized electrons. The molecular structures are based on the carbon sp^2 hybrid orbitals that make the sigma bonds and the p_z orbitals that combine with those of adjacent carbons to form the π bonds. The π bonds can connect with the π bonds of neighbouring molecules to form energy bands from the perspective of solid-state physics, or participate in intermolecular interactions, usually called π–π interactions, which are of particular interest to chemists.

The π–π interactions are prevalent in chemical and biological systems, and they have been under intense investigation in the last decade, having been the theme of a special issue of Chemistry of Materials [11], and another of Crystal Growth and Design [12]. In spite of their importance, they have no unified description, nor is there a consensus on the name used to describe this kind of interaction. They are known as the π–π interaction, π-stacking, etc. and there is even a proposal that such terms are misleading and should not be used [13]. In fact, contrary to the idea that the name may suggest, if one considers the intermolecular interaction between two benzene rings, it is found that a parallel stacked arrangement between aromatic rings does not exist, and that offset-stacked and face-to-edge orientations are the most favourable orientations for aromatic molecules [14]. The interactions involving π-electron systems are generally considered to be a combination of local or even atom-to-atom electrostatic and van der Waals interactions [3, 15–18] and, of course, steric effects [19].

This type of interaction can be described by the Hunter–Sanders model [3], according to which the major interaction energy comes from atom–atom electrostatic and van der Waals contributions. The contribution of the van der Waals interaction between two molecules can be calculated using equation (3.16). The electrostatic component is based on the idea of the 'π-system' [16], a simplification of the previous model based on local quadrupole interactions between aromatic rings. The simplest π-system is an sp^2 hybrid orbital set which provides three σ-bonds; a p_z orbital perpendicular to it forms the π-bond. The quadrupole is formed by two positive charges at the opposite sides of the σ framework of an aromatic ring, sandwiched between two regions of negatively charged π-electron density. The energetically favourable arrangements of the rings are (i) an edge-to-face or T-shaped geometry, i.e. a perpendicular arrangement of aromatic rings; and (ii) a parallel offset π-stacking geometry. Face-to-face stacking is energetically unfavourable due to the repulsion between the π-electron clouds. Most conjugated molecules with high carrier mobilities, such as pentacene in thin films, pack in a herringbone arrangement with tilted edge-to-face π–π interactions.

An interesting situation is that of parallel π-stacking interactions between planar radicals, which are much stronger than those between closed-shell rings. Quantum chemical calculations show that in the dimers of radicals there is a significant covalent character in the multicentre C–C bonds connecting the two molecules, called 'pancake bonding' [20]. This type of π-stacking interaction is also treated as a case of π-stacking interaction between non-aromatic rings [17]. A good example of multicentre covalent bonding is found in $[MeNC_5H_5]_2[TCNE]_2$ (TCNE=tetracyanoethylene), where in the dimer $TCNE^{-2}$, such C–C covalent bonds are 2.8Å long, as observed in x-ray and neutron diffraction data [21] (note that the van der Waals radius for C is 1.7 Å).

The theoretical methods more often used to calculate these interactions are based on DFT but also take dispersion interactions into account, as well as some implementations of symmetry-adapted perturbation theory (SAPT).

To conclude this section it is worth mentioning Roald Hoffmann's paper 'Interaction of Orbitals through Space and through Bonds' [22] as, in his words, a 'necessary addendum to the chemists view of a molecule. Localized orbitals or groups of orbitals may interact with each other directly, through space, or indirectly, through other bonds in the molecule.'

Further reading

Feynman R P, Leighton R B, and Sands M 2010 *The Feynman Lectures on Physics* vol 2 *Mainly Electromagnetism and Matter* (New Millennium Edition) (New York: Basic Books) chapter 6

Israelachvili J N 2011 *Intermolecular and Surface Forces* 3rd ed (Amsterdam: Elsevier)

Kaplan I G 2006 *Intermolecular Interactions: Physical Picture, Computational Methods and Model Potentials* (Chichester: Wiley)

Kleman M and Lavrentovich O D 2003 *Soft Matter Physics. An Introduction* (New York: Springer) chapter 1

Reifenberger R 2016 *Fundamentals of Atomic Force Microscopy: Part I: Foundations* (Singapore: World Scientific)

Problems

3.1 Derive Gauss's law (3.1) from Coulomb's law (3.4).

3.2 Consider the Lennard-Jones interatomic potential between two argon (Ar) atoms, assuming $w_0 = 0.238$ kcal mol^{-1} and $\sigma = 0.34$ nm.
- (a) Calculate the interatomic potential when the two atoms are 0.45 nm apart.
- (b) Calculate the value of the potential at its minimum.
- (c) At what distance does the potential become repulsive?
- (d) Calculate the van der Waals radius of argon.

References

[1] London F 1937 *Trans. Faraday Soc.* **33** 8b–26
[2] Pauling L 1945 *The Nature of the Chemical Bond* (Ithaca, NY: Cornell University Press)
[3] Hunter C A and Sanders J K M 1990 *J. Am. Chem. Soc.* **112** 5525–34
[4] Caillet J and Claverie P 1975 *Acta Cryst.* A **31** 448–61
[5] Arunan E *et al* 2011 *Pure Appl. Chem.* **83** 1637–41
[6] Steiner T *et al* 2002 *Angew. Chem. Int* **41** 48–76
[7] Glowacki E D, Irimia-Vladu M, Bauer S and Sariciftci N S 2013 *J. Mater. Chem.* B **1** 3742–53
[8] Modig K, Pfrommer B G and Halle B 2003 *Phys. Rev. Lett.* **90** 075502
[9] Israelachvili J N 2011 Special Interactions: Hydrogen-Bonding and Hydrophobic and Hydrophilic Interactions *Intermolecular and Surface Forces* 3rd ed (Amsterdam: Elsevier) 8
[10] Mahoney M W and Jorgensen W L 2000 *J. Chem. Phys.* **112** 8910–22
[11] 2011 *Chem. Mater.* **23** 309–922
[12] 2019 *Cryst. Growth Des.* Virtual Special Issue (https://pubs.acs.org/page/cgdefu/vi/20.html?ref=vi_collection)
[13] Martinez C R and Iverson B L 2012 *Chem. Sci.* **3** 2191–201
[14] Tran-Duc T, Thamwattana N, Cox B J and Hill J M 2010 *Philos. Mag.* **90** 1771–85
[15] Wheeler S E and Bloom J W G 2014 *J. Phys. Chem.* A **118** 6133–47
[16] Hunter C A, Lawson K R, Perkins J and Urch C J 2001 *J. Chem. Soc., Perkin Trans.* **2** 651–69
[17] Molčanov K and Kojić-Prodić B 2019 *IUCrJ* **6** 156–66
[18] Thakuria R, Nath N K and Saha B K 2019 *Cryst. Growth Des.* **19** 523–8
[19] Yao Z F, Wang J Y and Pei J 2018 *Cryst. Growth Des.* **18** 7–15
[20] Kertesz M 2019 *Chem. Eur. J.* **25** 400–416
[21] Piccoli P M B, Schultz A J, Sparkes H A, Howard J A K, Arif A M, Dawe L N and Miller J S 2009 *Cryst. Eng. Comm* **11** 686–90
[22] Hoffmann R 1971 *Acc. Chem. Res.* **4** 1–9

IOP Publishing

The Physics of Organic Electronics
From molecules to crystals and polymers
Luís Alcácer

Chapter 4

Structure and scattering

This chapter will briefly review the concepts of translational symmetry in crystals, direct and reciprocal lattices, periodic functions, Bravais lattices, and Bragg and diffuse scattering.

4.1 Introduction

Atoms and molecules have been found to form solids using many different packing arrangements according to their electronic structures and interactions. By looking at the periodic table of the elements, for at least the first six rows, we can generally predict the arrangements that are expected to be the most stable for the atoms of each element at low enough temperatures for them to assume equilibrial crystal structures. Let us start with the noble gases, He, Ne, etc. The atoms of these elements have very stable electronic structures. At very low temperatures their occupied energy levels are full, and the forces between atoms of the same species are of the weak van der Waals type. These elements, which (up to radon) are gases at room temperature, tend to form compact structures of the face-centred cubic (fcc) or hexagonal close-packed (hcp) types at low temperatures. In the first column of the periodic table, we find the *simple metals*, Li, Na, etc. Their atoms have one electron in the last shell, which they tend to lose, becoming ions with the electronic structure of a noble gas. The ions tend to arrange themselves in compact fcc crystal structures, and the free electrons tend to form an electron gas. In the second column we find the divalent metals, which tend to follow a similar rule, now involving two electrons. On the left of the noble gases we find F, Cl, etc., which tend to stabilize by capturing an electron to fill their last occupied shell. In this situation they tend to form chemical bonds that share electrons and they form diatomic molecules which interact via van der Waals forces. These elements tend to combine with elements of the simple metals, capturing one electron of the metal and forming ionic bonds which are electrostatic in nature. The crystal structures tend to be a combination of two

doi:10.1088/978-0-7503-3347-4ch4

interpenetrated cubic structures, satisfying electrical neutrality. A completely different situation is that of the so-called covalent solids, such as diamond (C), silicon (Si), and germanium (Ge). The atoms of these elements have electronic structures of the type [...ns^2 np^2]. The tendency is to redistribute the four valence electrons to form sp^3 hybrid orbitals, directed from the centre of a tetrahedron to the four vertex corners. The combination of these hybrid orbitals with those of similar neighbouring atoms forms typical covalent bonds. These are short, strong, and directed bonds. In diamond the C–C bond length is 1.54 Å (note that the van der Waals radius of carbon is 1.7 Å), and each carbon atom is linked to four other carbon atoms. Silicon and germanium have similar structures, with increasing bond lengths as we go down the column. The solids of these elements have fcc structures, but with tetrahedral primitive cells due to the geometry of the sp^3 orbitals. The electronic structure of these solids is characterized by the existence of two energy bands separated by a gap, resulting from the bonding and antibonding combination of the sp^3 orbitals. The gap, which is 5.5 eV for diamond, decreases to 1.14 eV for silicon and 0.67 eV for germanium. Tin, in the next position down the column, has two allotropes, one of which has the same structure, but its energy bands overlap slightly at the Fermi level and it is a semimetal. Consider solids such as gallium arsenide (GaAs), which is formed by an element to the left of Ge and an element to the right. These elements are the III–V semiconductors, because they are formed by an element of group III and an element of group V. The element of group III gets one electron from the element of group V; both become ions with the same electronic structure ($n\ sp^3$) and form the so-called zinc blende structure, which is similar to that of the group IV elements such as diamond and silicon, but now with the two different ions in the tetrahedral arrangement. We could go on and discuss these tendencies in more detail, but there are fundamental topics of crystalline materials which may be more relevant for our purpose, and that we should review.

4.2 Bravais lattices and crystal structures

Symmetry in physics and mathematics generally means *invariance*, denoting lack of change under some transformations, such as reflections, rotations, translations, etc. Symmetry is one of the most powerful and useful concepts in physics, and most of its power derives from the fact that the symmetry transformations of any system form a *group*, an elegant and simple algebraic structure of abstract algebra, which is defined as follows:

A *group* G is a set of elements $\{g_i\}$ that satisfy the following axioms:

1. If g_1 and g_2 belong to G, then the product $g_3 = g_1\, g_2$ also belongs to G.
2. Associativity: if g_1, g_2, and g_3 belong to G, then $g_1(g_2\, g_3) = (g_1\, g_2)g_3$.
3. Identity: there is an element $e \in G$, such that $eg = g$ for all $g \in G$.
4. Inverse: for any $g \in G$, there is an element g^{-1}, such that $g^{-1}\, g = e$.

The element e is unique; also, for each g, its inverse is unique. If $g_i\, g_j = g_j\, g_i$, the product is commutative and the group is called *abelian*; otherwise it is non-abelian.

Crystals are highly ordered atomic or molecular systems with low symmetry. Their regular arrangements have lower symmetry than liquids or gases. This may look counterintuitive at first sight, but consider, for example, how the density, $n(\mathbf{r})$, (number of particles per unit volume at point $\mathbf{r}$) changes in a gas–solid phase transition. In the gas phase the density is homogeneous, i.e. the density is the same everywhere, $n(\mathbf{r}) = n(\mathbf{r}')$ for any $\mathbf{r}$ and $\mathbf{r}'$: the density is invariant under any translation. In the solid (crystalline) phase the density is the same only at points with similar atomic neighbourhoods within the crystal, which are related by some particular translations represented by a set of vectors $\mathbf{T}$, such that $n(\mathbf{r}) = n(\mathbf{r} + \mathbf{T})$. We say that the density is invariant under all symmetry translations represented by $\mathbf{T}$, and that the phase transition is associated with a symmetry breaking—the symmetry decreases when going through the gas–solid phase transition.

This leads us to introduce the concepts of *translational symmetry* and the *Bravais lattice*. A Bravais lattice is an infinite array of discrete points generated by a discrete set of translation symmetry operators represented by vectors $\mathbf{T}$, which in three dimensions take the form

$$\mathbf{T} = n_1\mathbf{a} + n_2\mathbf{b} + n_3\mathbf{c} \tag{4.1}$$

where the basis vectors, $\mathbf{a}$, $\mathbf{b}$, and $\mathbf{c}$ are any three independent translational vectors which lie in different directions and span the lattice, and n_1, n_2, and n_3 are integers. $\mathbf{T}$ represents a symmetry translation (or simply a translation), i.e. a shift under which the lattice is invariant. Figure 4.1 illustrates a symmetry translation for a two-dimensional Bravais lattice where $\mathbf{T} = 2\mathbf{a} + \mathbf{b}$, which relates two points $\mathbf{r}$ and $\mathbf{r}'$ with identical environments, such that $\mathbf{r}' = \mathbf{r} + \mathbf{T}$. All points of a 3D Bravais lattice are defined by linear combinations of the basis $\mathbf{a}$, $\mathbf{b}$, and $\mathbf{c}$, with integer coefficients. The $\mathbf{a}$, $\mathbf{b}$, and $\mathbf{c}$ vectors define a *unit cell*, which is a parallelepiped with a volume of $V_c = |\mathbf{a} \cdot (\mathbf{b} \times \mathbf{c})|$. The angles opposite to $\mathbf{a}$, $\mathbf{b}$, and $\mathbf{c}$ are called α, β, and γ, respectively. The smallest unit cell is the *primitive cell*. A primitive cell, with the lattice point in the middle, is called the Wigner–Seitz cell, which is the smaller volume delimited by bisector planes perpendicular to all lattice vectors emerging from a given lattice point.

A Bravais lattice with the atoms in their equilibrium positions forms a *crystal structure*. The atoms and their positions in the unit cell defined in terms of the basis vectors $\mathbf{a}$, $\mathbf{b}$, and $\mathbf{c}$ form the *basis of atoms*.

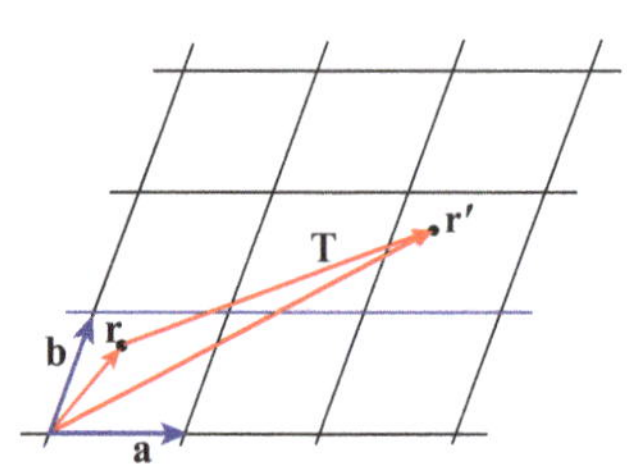

Figure 4.1. A two-dimensional Bravais lattice. $\mathbf{T} = \mathbf{r}' - \mathbf{r}$ is a (symmetry) translation vector.

Bravais lattices and crystal structures can be classified according to their symmetry group. In three dimensions, there are 14 Bravais lattices, 7 crystal systems, 32 crystallographic point groups, and 230 space groups. In two dimensions, there are 5 Bravais lattices, 4 crystal systems, 10 crystallographic point groups and 17 space groups. Let us try to explain what this means.

Imagine a cube with an axis going through the centre of the cube and the middle points of two opposite faces. A rotation of the cube around the axis by an angle of 90° would lead to a similar view—the cube is invariant under a 90° rotation around that axis. In the context of symmetry there are two different entities to take into account: the *symmetry elements*, as the axes, and the *symmetry operations*, as the rotations. The symmetry elements are the points, lines, and planes relative to which we can perform symmetry operations. Symmetry operations are: rotations (of $2\pi/n$, denoted by C_n with n an integer), reflections relative to a symmetry plane (denoted by σ), inversion (denoted by i, under which all coordinates, x, y, z, change to $-x$, $-y$, $-z$), improper rotations (rotations of $2\pi/n$ followed by a reflection, denoted by S_n) and, of course, the *identity* operation (denoted by E) which consists of doing nothing.

It would be easy to check that all the symmetry operations of an object, such as the cube, form a group in the sense of group theory, i.e. they satisfy the four axioms of a group. For example, we can define the product of two rotations of 90°as a rotation of 180°; we can do one rotation and later on two more rotations, which would be the same as doing two rotations and another one later on, but in the same order (associativity); we can do a rotation clockwise and then a rotation counter-clockwise by the same angle, which would take the cube to the initial position (inversion); or we could do nothing—just look—(identity).

The symmetry operations that keep one point fixed are called *point operations* and the corresponding group is a *point group*. Note that all possible rotations that keep a view of an object invariant themselves form a group which is a subgroup of the group of all the symmetry operations of that object. For a sphere, we can consider rotations around any axis going through the centre by any angle, and reflections about any plane passing through the centre.

To use these concepts we need mathematical representations of the symmetry operations. We may use a collection of matrices that form a group that is *isomorphic* with the original group, meaning that there is a one-to-one correspondence between the elements of the two groups. The theory of group representations is not only very useful but also full of mathematical beauty (as Dirac would say), but we will not go into it.

In crystals, in addition to the symmetry point groups, which preserve a fixed point, there is the group of symmetry translations, or the translation group. The set of all symmetry operations (point operations combined with translations) of a crystal structure is called a *space group*. The symmetry group of a Bravais lattice is the translation group represented by the vectors $\mathbf{T}$ (equation (4.1)) together with some discrete point group about any lattice point. Exceptionally, $\mathbf{T}$ can be fractional, if there is a screw rotation or a glide reflection.

The Bravais lattices are the following: (1) cubic P (simple); (2) cubic I (body centred); (3) cubic F (face centred); (4) tetragonal P (simple); (5) tetragonal I (body centred); (6) orthorhombic P; (7) orthorhombic I; (8) orthorhombic C; (9) orthorhombic F; (10) monoclinic P; (11) monoclinic C; (12) triclinic; (13) trigonal R; and (14) hexagonal P.

The seven crystal systems correspond to Bravais lattices with the same point group as the crystal: (1) O_h (octahedral, or cubic); (2) D_{4h} (tetragonal); (3) D_{2h} (orthorhombic); (4) C_{2h} (monoclinic); (5) C_i (triclinic); (6) D_{3d} (trigonal); (7) D_{6h} (hexagonal). O_h, D_{4h}, etc. are the point groups (also the symmetry groups of molecules).

Any (3D) crystal belongs to one of the 230 space groups (crystal symmetry groups), one of the 32 point groups, one of the 14 Bravais lattices, and one of the 7 crystal systems.

4.2.1 The reciprocal space, the reciprocal lattice, and the Brillouin zone

Photons, electrons, and other particles, which behave as waves when they propagate within a crystal, can be scattered due to the periodicity of the crystal lattice. To understand this phenomenon, we need to know how these waves interact with the atoms in crystals.

Consider a plane wave $\exp(\mathrm{i}\,\mathbf{k}\cdot\mathbf{r})$ with an incidence perpendicular to a set of parallel lattice planes equally separated by a distance a. For an arbitrary value of $k = |\mathbf{k}|$, the wave with wavelength $\lambda = 2\pi/k$ will not have the periodicity of the lattice except when a is a multiple of the wavelength, i.e. for the values that satisfy the condition $\lambda = a/n$ or $k = 2\pi\, n/a$, with n an integer. When $\mathbf{a}$, $\mathbf{b}$, and $\mathbf{c}$ are orthogonal, for example, along x, y, and z, the vectors representing waves with the periodicity of the lattice are of the form $\mathbf{k}_1 = (2\pi/a)n_1\mathbf{e}_x$, $\mathbf{k}_2 = (2\pi/b)n_2\mathbf{e}_y$, and $\mathbf{k}_3 = (2\pi/c)n_3\mathbf{e}_z$, respectively. This suggests the definition of a wave vector space, or *reciprocal space*, with the basis $\mathbf{g}_1 = (2\pi/a)\mathbf{e}_x$, $\mathbf{g}_2 = (2\pi/b)\mathbf{e}_y$ and $\mathbf{g}_3 = (2\pi/c)\mathbf{e}_z$. If $\mathbf{a}$, $\mathbf{b}$, and $\mathbf{c}$ are non-orthogonal, the basis of the reciprocal space can be written as

$$\begin{aligned} \mathbf{g}_1 &= 2\pi\frac{\mathbf{b}\times\mathbf{c}}{\mathbf{a}\cdot(\mathbf{b}\times\mathbf{c})} \\ \mathbf{g}_2 &= 2\pi\frac{\mathbf{c}\times\mathbf{a}}{\mathbf{a}\cdot(\mathbf{b}\times\mathbf{c})} \\ \mathbf{g}_3 &= 2\pi\frac{\mathbf{a}\times\mathbf{b}}{\mathbf{a}\cdot(\mathbf{b}\times\mathbf{c})} \end{aligned} \tag{4.2}$$

where $\mathbf{a}\cdot(\mathbf{b}\times\mathbf{c})$ is the volume, V_c, of the unit cell in direct space.

A vector of the form

$$\mathbf{G} = h\,\mathbf{g}_1 + k\,\mathbf{g}_2 + l\,\mathbf{g}_3 \qquad h,k,l = \text{integers} \tag{4.3}$$

generates a new Bravais lattice, called the *reciprocal lattice*.

Note that we will always have

$$e^{\mathrm{i}\,\mathbf{G}\cdot(\mathbf{r}+\mathbf{T})} = e^{\mathrm{i}\,\mathbf{G}\cdot\mathbf{r}} \tag{4.4}$$

since the plane waves $\exp(\mathrm{i}\,\mathbf{G}\cdot\mathbf{r})$ have the periodicity of the lattice. By dividing both sides of equation (4.4) by the common factor $\exp(\mathrm{i}\,\mathbf{G}\cdot\mathbf{r})$, we can define the reciprocal lattice as the set of wave vectors **G** which satisfy the condition

$$e^{\mathrm{i}\,\mathbf{G}\cdot\mathbf{T}} = 1. \tag{4.5}$$

A reciprocal lattice is defined relative to a particular Bravais lattice in direct space, which we may call a *direct lattice*, when considered in relation to its reciprocal. Equation (4.4) can be generalized for any function with the periodicity of the lattice. In fact, the translational symmetry implies that, for any $f(\mathbf{r})$ with the periodicity of the lattice,

$$f(\mathbf{r}+\mathbf{T}) = f(\mathbf{r}) \tag{4.6}$$

for all points **r** and for all vectors **T**. We can, therefore, write $f(\mathbf{r})$ as a Fourier series:

$$\begin{aligned} f(\mathbf{r}) &= \sum_{\mathbf{G}} f_{\mathbf{G}}\, e^{\mathrm{i}\,\mathbf{G}\cdot\mathbf{r}} \\ f_{\mathbf{G}} &= \frac{1}{V_c}\int_c f(\mathbf{r})\, e^{-\mathrm{i}\,\mathbf{G}\cdot\mathbf{r}}\, d\mathbf{r} \end{aligned} \tag{4.7}$$

where V_c is the volume of the unit cell and $d\mathbf{r}$ the volume element.

The study of periodic functions with the periodicity of a crystal lattice is of great importance in solid-state physics. In fact, many physical quantities, such as the scattering potential $V(\mathbf{r})$, the electron density $\rho(\mathbf{r})$, and the number density $n(\mathbf{r})$, have the periodicity of the crystal lattice and can be written in the form of equation (4.7).

From (4.1), (4.3), and (4.2), we can conclude that the product $\mathbf{G}\cdot\mathbf{T}$ can be written as

$$\mathbf{G}\cdot\mathbf{T} = (h\,n_1 + k\,n_2 + l\,n_3)2\pi = 2\pi\, m \qquad m = \text{integer} \tag{4.8}$$

which confirms (4.5) and implies

$$\mathbf{g}_i\cdot\boldsymbol{\tau}_j = 2\pi\,\delta_{ij} \qquad \boldsymbol{\tau}_j = \mathbf{a}, \mathbf{b}, \mathbf{c} \quad \delta_{ij} = \begin{cases} 1 & \text{if } i = j \\ 0 & \text{if } i \neq j \end{cases}.$$

This also means that $\mathbf{g}_i$ are perpendicular to $\boldsymbol{\tau}_j$ when $j \neq i$.

As in direct space, we define a unit cell in reciprocal space as the volume delimited by the $\mathbf{g}_1$, $\mathbf{g}_2$, and $\mathbf{g}_3$ vectors (assuming we are in three dimensions), and we can define the Wigner–Seitz cell of a reciprocal lattice which is called the *first Brillouin zone* or simply the *Brillouin zone*. Note that the units of reciprocal space are L^{-1} (e.g. m^{-1}, $Å^{-1}$). The lattice parameters a, b, and c for a typical crystal lattice are of the order of the Å, while the corresponding reciprocal parameters g_1, g_2, and g_3 are of the order of $Å^{-1}$ ($10^{10}m^{-1}$), and these are, in fact, *wave numbers*.

The integers h,k,l are called the Miller indices and they define the lattice planes and directions in the lattice. The crystal planes are defined by the inverse of the intersections with the **a**, **b**, and **c** axes, $(a/h,\ b/k,\ c/l)$, in that order, after suppressing common factors. Since they are ordered, we just need to use the indices $(h\,k\,l)$ to define the planes. The *director* (or normal) of a plane ($h\,k\,l$) is the vector $\mathbf{G}_{hkl}$. Note that the equation of a plane can be written as $x/x_0 + y/y_0 + z/z_0 = 1$, where x_0, y_0, and z_0 are the intersections with the respective axis. Therefore, we can write the equation for the plane ($h\,k\,l$) as $hx + ky + lz = 1$. The conventional notation is as follows: ($h\,k\,l$) for a plane; (h, k, l) for a point; [$h\,k\,l$] for a direction; <$h\,k\,l$> for a family (a family is a symmetry related set) of directions and $\{h\,k\,l\}$ for a family of planes. Negative Miller indices are denoted by a minus sign over the index. For example, in a face-centred cubic lattice, the plane (200) intersects the **a**-axis at $a/2$ and is parallel to **b** and **c** (it intersects those axes at infinity, i.e. $b/0 = \infty$, $c/0 = \infty$); the plane (110) intersects the **a** and **b** axes at a and b, and is parallel to **c**. Another example, the ($1\bar{1}1$) plane belongs to the $\{111\}$ family (of symmetry related planes).

Any vector **G** is perpendicular to a set of planes in the direct lattice. Considering that the projection of **T** onto **G** is $\mathbf{G}\cdot\mathbf{T} = |\mathbf{G}||\mathbf{T}|\cos\theta$ (figure 4.2(b)) and that the distance to the origin of the plane normal to **G** and passing through **r** is $d = |\mathbf{T}|\cos\theta$, we have $d = \mathbf{G}\cdot\mathbf{T}/|\mathbf{G}|$.

As we can see from figure 4.2, an infinite number of vectors such as **T** and **T**′ end in points of the (10) planes. And we can imagine that similar situations occur in three-dimensional lattices. Thus, in general, if there is one atom at a given position in a given crystal plane, there will be a large number in that plane. Note that the lattices, as defined, are infinite, although real crystals are finite, but have a large number of atoms, with the exception of nanocrystals. In this way, we can build a crystal plane whose director is $\mathbf{G}_{hkl}$. On the other hand, all **T**s that satisfy (4.8) for a given **G** end up in a plane normal to **G**. Then each **G**, or its ($h\,k\,l$) indices, can be used to refer to any set of crystal planes, and, since these indices are defined within a common factor, we can choose the set of minimum integer indices ($h\,k\,l$), such that **G** is the shorter translation vector of the reciprocal lattice. This shows that the distance between adjacent crystal planes defined by ($h\,k\,l$) is given by

$$d_{hkl} = \frac{2\pi}{|\mathbf{G}_{hkl}|} \tag{4.9}$$

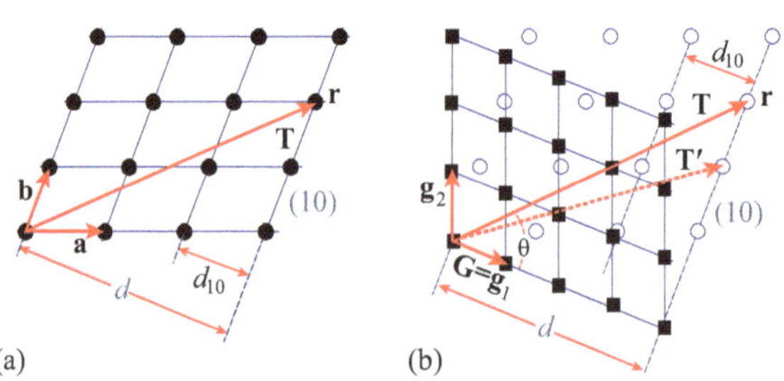

(a) (b)

Figure 4.2. (a) A two-dimensional lattice (circles) with a point **r** related to the origin by a translation vector **T**. Point **r** lies in a (10) line which is at a distance d from the origin; the distance d_{10} between adjacent (10) planes is also shown. (b) The associated reciprocal lattice (squares) superposed on the direct lattice (open circles), with the reciprocal lattice basis vectors and the $\mathbf{G} = \mathbf{g}_1$ vector which is normal to the (10) plane. The distance to the origin of the plane (10) passing through **r** is $d = |\mathbf{T}|\cos\theta$.

and the shortest $\mathbf{G}$ is $\mathbf{G}_0 = (2\pi/d)\mathbf{n}$, where $\mathbf{n}$ is the normal to the set of planes. $\mathbf{G}_{hkl}$ therefore defines the $[h\,k\,l]$ direction.

4.3 Scattering and diffraction

Late at night, on Friday, 8 November, 1895, alone in his private laboratory in the city of Würzburg in Germany, Wilhelm Roentgen certainly got a scare when he saw the ghostly skeleton of his own hand projected on a screen [1]. He had accidentally discovered x-rays.

Almost 18 years later, in 1913, W H and W L Bragg [2] discovered that the x-ray diffraction patterns produced by crystals were due to reflections of the x-rays in planes where the atoms are located within the crystal. The x-ray beams were reflected by those planes, separated by the interplanar distance d, each plane reflecting only a small fraction of the incident radiation (figure 4.3(a)). The reflected rays interfered constructively when the path difference, $2d \sin\theta$, was an integer multiple of the wavelength λ, since only then were the waves in phase; θ was the glancing angle. The Bragg condition is therefore

$$2d \sin\theta = n\,\lambda \tag{4.10}$$

where $n = 1$ for first-order reflection. The wavelength of the x-rays used is typically of the order of the Å, which is comparable to the atomic distances in solids. A beam of x-rays incident on a crystal is diffracted at angles which satisfy the Bragg condition (4.10) for the various crystal planes. Although x-ray diffraction is used to determine crystal structures, the scattering laws are relevant for many processes in condensed-matter physics and are applied in many instances that involve interactions between electromagnetic radiation or particles, such as neutrons or electrons, and matter in periodic structures.

4.3.1 X-ray diffraction

Let us consider a perfect crystal whose atoms are in their equilibrium positions, which naturally generates a periodic potential. If we hit such a crystal with a beam of x-rays with a wave vector $\mathbf{k}$, there is a probability amplitude that the beam will scatter into a wave vector $\mathbf{k}'$ proportional to $\langle \mathbf{k}'|V(\mathbf{r})|\mathbf{k}\rangle$, where $V(\mathbf{r})$ is the interaction

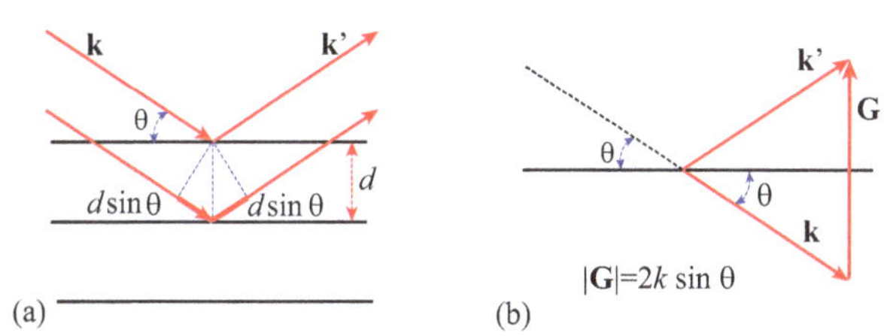

Figure 4.3. (a) Diffraction due to parallel planes, illustrating Bragg's law. The magnitude of both $\mathbf{k}$ and $\mathbf{k}'$ is $2\pi/\lambda$, and the path difference between waves partially reflected from successive planes is $2d \sin\theta$, d being the interplanar distance. (b) Laue's condition implies that whenever the relation $\Delta\mathbf{k} = \mathbf{k}' - \mathbf{k} = \mathbf{G}$ is satisfied, constructive interference takes place. As shown in the figure, the two relations $|\mathbf{G}| = 2k \sin\theta$ and $d = 2\pi/|\mathbf{G}|$ are equivalent.

potential. We can expand $V(\mathbf{r})$ in a Fourier series according to (4.7), $V(\mathbf{r}) = \sum_{\mathbf{G}} V_{\mathbf{G}}\, e^{i\,\mathbf{G}\cdot\mathbf{r}}$, where $\mathbf{G}$ is a reciprocal lattice vector. Considering that $|\mathbf{k}\rangle = e^{i\,\mathbf{k}\cdot\mathbf{r}}$ and $|\mathbf{k}'\rangle = e^{i\,\mathbf{k}'\cdot\mathbf{r}}$ are the incident and scattered waves, respectively (leaving out normalization factors), the scattering amplitude is of the form

$$\langle\mathbf{k}'|V|\mathbf{k}\rangle = \int e^{-i\,\mathbf{k}'\cdot\mathbf{r}} V(\mathbf{r}) e^{i\,\mathbf{k}\cdot\mathbf{r}} d\mathbf{r} = \sum_{\mathbf{G}} \int e^{-i\,\mathbf{k}'\cdot\mathbf{r}} V_{\mathbf{G}}\, e^{i\,\mathbf{G}\cdot\mathbf{r}} e^{i\,\mathbf{k}\cdot\mathbf{r}} d\mathbf{r} = \sum_{\mathbf{G}} \int V_{\mathbf{G}} e^{i\,(\mathbf{k}+\mathbf{G}-\mathbf{k}')\cdot\mathbf{r}} d\mathbf{r}$$

which only will be non-zero if $\mathbf{k} + \mathbf{G} - \mathbf{k}' = 0$, i.e. if

$$\Delta\mathbf{k} = \mathbf{k}' - \mathbf{k} = \mathbf{G}. \tag{4.11}$$

This is *Laue's condition*. Considering (4.9) and (4.11), which imply $|\mathbf{G}| = 2k \sin\theta$, as shown in figure 4.3(b), and that $k = 2\pi/\lambda$, we recover Bragg's law, $2\,d \sin\theta = \lambda$.

The scattering potential depends on the positions of the nuclei and on the distribution of the electrons around them. In general, the atoms are well separated and we may neglect the overlap of the electron clouds of neighbouring atoms. We may then consider the scattering potential of a crystal to be a superposition of the contributions from the individual atoms at positions $\mathbf{R}_j$ in the crystal, i.e. $V(\mathbf{r}) = \sum_j^N v_j(\mathbf{r} - \mathbf{R}_j)$. When condition (4.11) is satisfied, diffraction occurs with an intensity proportional to $|A(\mathbf{G})|^2$, where $A(\mathbf{G})$ is the amplitude:

$$A(\mathbf{G}) = \langle\mathbf{k}'|V|\mathbf{k}\rangle = \int e^{-i\,\mathbf{k}'\cdot\mathbf{r}} V(\mathbf{r}) e^{i\,\mathbf{k}\cdot\mathbf{r}} d\mathbf{r} = \int e^{-i\,\mathbf{G}\cdot\mathbf{r}} \sum_j^N v(\mathbf{r} - \mathbf{R}_j)\, d\mathbf{r}.$$

The sum is taken over all N atoms in the crystal, and the integral is taken over the volume of the crystal.

Using the translation invariance of the lattice, we can represent any point $\mathbf{r}$ in the crystal as $\mathbf{r} = \mathbf{T}_n + \mathbf{r}'$, where $\mathbf{T}_n$ is a translation vector (4.1) with n denoting the set of integers $n = (n_1, n_2, n_3)$, and $\mathbf{r}'$ is the same point in the coordinate system of its unit cell (figure 4.1). We can then write the scattering amplitude as

$$A(\mathbf{G}) = \left(\int \sum_j^s e^{-i\,\mathbf{G}\cdot\mathbf{r}'} v(\mathbf{r}' - \mathbf{r}_j)\, d\mathbf{r}'\right) \sum_n e^{-i\,\mathbf{G}\cdot\mathbf{T}_n}$$

where the parentheses relate to the unit cell, i.e. the integral is taken over its volume, the sum is taken over the s atoms in the cell, and $\mathbf{r}_j$ is the position of the jth atom in the coordinate system of the unit cell. The sum over n extends to all translation vectors in the crystal.

To get the desired result we multiply and divide by $\exp(i\,\mathbf{G}\cdot\mathbf{r}_j)$, and rearrange:

$$\begin{aligned} A(\mathbf{G}) &= \left(\sum_j^s \int e^{-i\,\mathbf{G}\cdot(\mathbf{r}'-\mathbf{r}_j)} v(\mathbf{r}' - \mathbf{r}_j)\, e^{-i\,\mathbf{G}\cdot\mathbf{r}_j}\, d\mathbf{r}'\right) \sum_n e^{-i\,\mathbf{G}\cdot\mathbf{T}_n} \\ &= \left(\sum_j^s f_j\, e^{-i\,\mathbf{G}\cdot\mathbf{r}_j}\right) \sum_n e^{-i\,\mathbf{G}\cdot\mathbf{T}_n} = F(\mathbf{G})\,\Sigma(\mathbf{G}) \qquad n = (n_1, n_2, n_3) \end{aligned} \tag{4.12}$$

where

$$f_j = \int e^{-i\,\mathbf{G}\cdot(\mathbf{r}'-\mathbf{r}_j)}\, v(\mathbf{r}' - \mathbf{r}_j)\, d\mathbf{r}'. \tag{4.13}$$

In (4.12), the expression inside the parentheses, $F(\mathbf{G})$, is the *structure factor of the unit cell*, and $\Sigma(\mathbf{G})$ is called the *form factor of the crystal*; f_j (4.13) is the *atomic form factor* or *atomic scattering factor*.

The structure factor $F(\mathbf{G})$ is calculated by using equation (4.3) to specify $\mathbf{G}_{hkl}$ and the positions of the atoms in terms of its components in the unit cell, i.e. $\mathbf{r}_j = u_j\mathbf{a} + v_j\mathbf{b} + w_j\mathbf{c}$:

$$F_{\mathrm{G}} = \sum_{j=1}^{s} f_j\, e^{-i(\mathbf{G}\cdot\mathbf{r}_j)} = \sum_{j=1}^{s} f_j\, e^{-i2\pi(u_jh+v_jk+w_jl)}. \tag{4.14}$$

Scattering phenomena are generally characterized by the differential scattering cross section, $d\sigma/d\Omega$, defined as the ratio of the number of particles scattered per unit time per unit solid angle $d\Omega$, which is proportional to the time average of $|A(\mathbf{G})|^2$, where the electron density $\rho(\mathbf{r}, t)$ at the position $\mathbf{r}$ and time t is used instead of the atomic scattering potential $v(\mathbf{r}' - \mathbf{r}_j)$. When Laue's condition (4.11) is fulfilled, all unit cells scatter in phase, and the diffracted x-rays give rise to Bragg spots on 2D detectors whose intensity is proportional to $|A(\mathbf{G})|^2$. Structure determination methods are based on iterative procedures that start from initial models, from which structure factors are calculated and compared with those observed experimentally.

Structure factors are widely used in condensed-matter physics in different versions of equation (4.12) in which the potential $v_j(\mathbf{r}_j)$ in the form factor (4.13) is replaced by appropriate functions, such as the spatial density, $\rho(\mathbf{r})$, the number density $n(\mathbf{r})$, or the density–density correlation function $C_{nn}(\mathbf{r}, \mathbf{r}')$.

4.3.2 Diffuse scattering

Diffuse scattering arises from any departure from long-range order in a material's structure. While the Fourier transform of an ordered structure gives well-defined spots corresponding to Bragg reflections, the Fourier transform of a disordered structure gives a spread of scattering intensities over a wide range of $\boldsymbol{\Delta}\mathbf{k}$ values. In a review of x-ray diffuse scattering in molecular conductors, Sylvain Ravy [3] explained how this method was successfully used to study phase transitions and structural instabilities in organic conductors, and gave many examples, namely on the Bechgaard salts, including $(TMTSF)_2PF_6$, which will be one of our case studies in chapter 6. Here, we will give a very interesting example of how diffuse scattering has contributed to the understanding of the spin Peierls instability in one of the perylene-dithiolate salts (the other case study in chapter 6).

The scattering amplitude for diffuse scattering is given by

$$A(\mathbf{G}) = \sum_{n} F_n(\mathbf{G}) e^{-i\,\mathbf{G}\cdot\mathbf{T}_n} \qquad n = (n_1, n_2, n_3),$$

where $F_n(\mathbf{G})$ is a structure factor which depends on the instantaneous content of the unit cell located at $\mathbf{T}_n = n_1\mathbf{a} + n_2\mathbf{b} + n_3\mathbf{c}$, and that can be written as

$$F_n(\mathbf{G}) = \langle F_n(\mathbf{G})\rangle_{n,t} + \phi_n(\mathbf{G}), \tag{4.15}$$

where $\langle F_n(\mathbf{G})\rangle_{n,t}$ is the spatial and time average, to be calculated in that order, and $\phi_n(\mathbf{G})$ refers to the diffuse scattering.

Using equation (4.15) in (4.12) implies that the scattered intensity is related to the Fourier transform of the pair correlation function:

$$\langle F_n^*(\mathbf{G})\, F_{n+m}(\mathbf{G})\rangle_{n,t} = \left|\langle F_n(\mathbf{G})\rangle_{n,t}\right|^2 + \langle \phi_n^*(\mathbf{G})\, \phi_{n+m}(\mathbf{G})\rangle_{n,t}. \tag{4.16}$$

The first term on the right-hand side of equation (4.16) is the diffraction contribution. It gives the intensity of the Bragg spots, and is due to the long-range order in the crystal. The presence of disorder (e.g. thermal agitation of the atoms) in the diffraction contribution is included in the structure factor average $\langle F(\mathbf{G})\rangle$. The second term on the right-hand side of the equation is the *diffuse scattering* contribution, and is due to the existence of *short-range order* (SRO) in the crystal. Diffuse scattering is not localized at the (h, k, l) points of the reciprocal lattice but spread out in the wave vector and for that reason is also called out-of-Bragg scattering. Diffuse scattering can be concentrated in diffuse rings, diffuse sheets, or diffuse lines. The intensity of the diffuse scattering patterns $I_{DD}(\mathbf{G})$ is given by

$$I_{DD}(\mathbf{G}) = N\sum_m \langle \phi_n^*(\mathbf{G})\, \phi_{n+m}(\mathbf{G})\rangle_{n,t} e^{\mathrm{i}\,\mathbf{G}\cdot\mathbf{T}_m}.$$

This equation shows that only deviations from the perfect periodicity ($\phi_n \neq 0$) lead to diffuse scattering. Diffuse x-ray scattering is a powerful method for studying disorder in materials, and is very useful in the study of phase transitions and lattice instabilities in organic conductors. Figure 4.4 is an interesting illustration of the information that can be retrieved from x-ray diffuse scattering patterns. The compound $\mathrm{Per_2[Pt(mnt)_2]}$, described in detail in chapter 6, is a metal at room temperature and undergoes a metal–insulator transition at $\approx$8 K, the real nature of which is still controversial. Diffuse scattering measurements recently reported by Pouget *et al* [4] show that at 10 K (slightly above the temperature of the metal–insulator transition), intense diffuse lines appear at the $\mathbf{q} = \mathbf{b}^*/2$ wave vector (by convention, $\mathbf{b}\cdot\mathbf{b}^* = 2\pi$), between the layers of the Bragg spots, which are a signature of lattice fluctuations preceding dimerization of the quasi one-dimensional lattice. At 4 K, i.e. below the metal–insulator transition, these lines have condensed into broad diffuse spots, indicating the presence of SRO instead of the well-defined spots expected after dimerization of the lattice in the phase transition. The resulting range order, which was measured, extends over 36 Å($\approx 9b$) in the $\mathbf{b}$ direction and 16 Å in the transverse direction. These findings are valuable for establishing the nature of the phase transition.

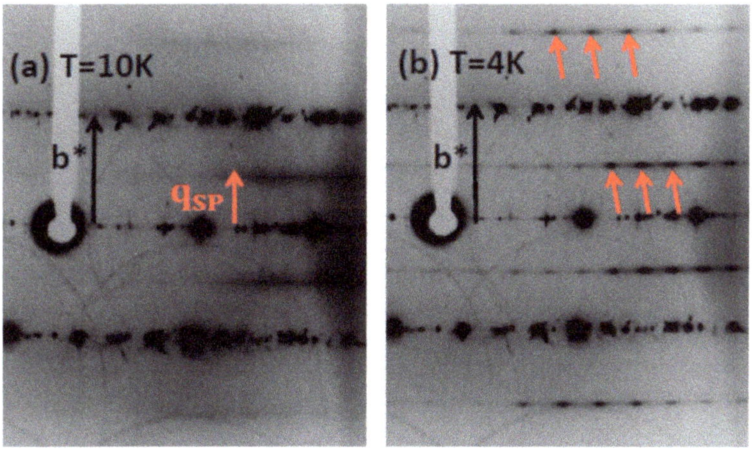

Figure 4.4. X-ray diffuse scattering patterns from $Per_2[Pt(mnt)_2]$ before (a), and after (b) the metal–insulator transition ($T_c \approx 8$ K). In (a) the arrow points to diffuse lines due to lattice fluctuations above T_c, located half way between the Bragg spots, i.e. at the wave vector $\mathbf{q} = \mathbf{b}^*/2$. In (b) the arrows show that the previous diffuse lines have condensed into broad diffuse spots, corresponding to the establishment of an SRO below T_c. Reprinted with permission from [4], copyright 2017 by the authors.

4.3.3 Electrons in a periodic potential

Let us now consider how the motion of electrons in a crystal is affected by a periodic potential. An electron with a wave vector $\mathbf{k}$ will be reflected in the $(h\,k\,l)$ planes when Laue's condition (4.11), $\mathbf{k}' = \mathbf{k} + \mathbf{G}_{hkl}$, is satisfied. For elastic scattering $|\mathbf{k}|^2 = |\mathbf{k}'|^2$, therefore

$$\mathbf{k}'^2 = (\mathbf{k} + \mathbf{G}_{hkl})^2 = \mathbf{k}^2 + 2\,\mathbf{k} \cdot \mathbf{G}_{hkl} + \mathbf{G}_{hkl}^2 = \mathbf{k}^2$$

which leads to

$$2\,\mathbf{k} \cdot \mathbf{G}_{hkl} + \mathbf{G}_{hkl}^2 = 0.$$

If $\mathbf{G}$ is a reciprocal lattice vector, then $-\mathbf{G}$ is as well, and the diffraction condition becomes

$$\mathbf{k} \cdot \frac{1}{2}\mathbf{G} = \left(\frac{1}{2}\mathbf{G}\right)^2. \tag{4.17}$$

The construction of Brillouin zones is difficult to follow for three-dimensional lattices, but easy for one- and two-dimensional ones. Let us then consider a one-dimensional lattice with N primitive cells of size a. The $\mathbf{G}$ vectors are of the form $\mathbf{G} = h\,\mathbf{g}_1$ with $\mathbf{g}_1 = (2\pi/a)\mathbf{e}_x$, and the Brillouin zone is delimited by the planes which bisect $\mathbf{g}_1$ and $-\mathbf{g}_1$, i.e. at $k = \pm\pi/a$, as can be verified using (4.17). This implies that waves with $k = \pm\pi/a$ will be reflected.

Another example is that of a square lattice. Let us draw the Brillouin zones (figure 4.5). The regions of reciprocal space delimited by the Bragg planes are the Brillouin zones. The first Brillouin zone, denoted by 1, is the Wigner–Seitz cell of the reciprocal lattice. The four triangles adjacent to the central square form the second Brillouin zone. Together they have the same area and can be folded, or 'flipped over'

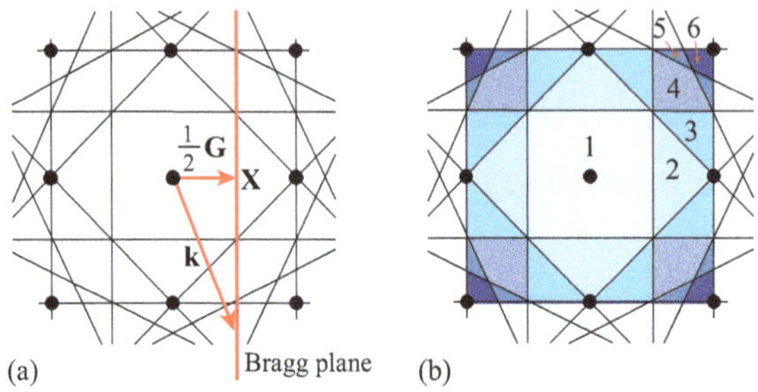

Figure 4.5. (a) Brillouin zone construction. Starting from a reciprocal lattice point, we imagine lines connecting the neighbouring points and draw planes which bisect them. These planes, which satisfy (4.17), are the zone boundaries (or Bragg planes). (b) The regions delimited by the Bragg planes are the Brillouin zones. The numbers denote their order.

the first Brillouin zone. This operation is called *reduction* to the (first) Brillouin zone. The following sets of regions all have the same area and can be folded over the first Brillouin zone (or simply Brillouin zone). Electrons with wave vectors **k** ending in the zone boundaries are scattered from the Bragg planes, the degree of scattering $\mathbf{k} \rightarrow \mathbf{k}'$ being proportional to $\langle \mathbf{k}' | V | \mathbf{k} \rangle$.

In a scattering process involving two particles (e.g. electron–electron, electron–phonon) for which the resulting wave vector $\mathbf{k}_1 + \mathbf{k}_2 = \mathbf{k}_3$ falls outside the first Brillouin zone, a reciprocal lattice vector is added, $\mathbf{k}_1 + \mathbf{k}_2 = \mathbf{k}_3 + \mathbf{G}$, so that the process can be represented in the first Brillouin zone. Such a process is called an *Umklapp* (flipped-over) process.

4.3.4 Consequences of translational invariance

In the approximation in which the ions are in their equilibrium positions, the lattice part of the periodic potential experienced by electrons in a crystal lattice, included in the Schrödinger equation, e.g. in the Hartree–Fock approximation, is invariant to all operations of the space group, in particular, to the group of translations, and we can use equation (4.6) for that potential, i.e. $V(\mathbf{r} + \mathbf{T}) = V(\mathbf{r})$. A theorem derived by Bloch [5] shows that the solutions of the Schrödinger equation can be chosen to be the products of plane waves, $\exp(\mathrm{i}\,\mathbf{k} \cdot \mathbf{r})$, and appropriate functions (e.g. orbitals in the unit cell), $u_{n\mathbf{k}}(\mathbf{r})$, with the periodicity of the lattice:

$$\begin{aligned} \psi_{n\mathbf{k}}(\mathbf{r}) &= u_{n\mathbf{k}}(\mathbf{r})\, e^{i\,\mathbf{k}\cdot\mathbf{r}} \\ u_{n\mathbf{k}}(\mathbf{r} + \mathbf{T}) &= u_{n\mathbf{k}}(\mathbf{r}). \end{aligned} \tag{4.18}$$

The $\psi_{n\mathbf{k}}(\mathbf{r})$ functions are called *Bloch functions; n* is a quantum number which represents each of the solutions of the Schrödinger equation, and the wave vector **k** has a similar role to that of free electrons in the Sommerfeld model, except that it is not proportional to the velocity of the electrons, and is called the *crystal momentum*. If we make u a constant in equation (4.18), the wave function ψ becomes a plane wave. This means that the Bloch functions are plane waves which are modulated by the periodicity of the lattice.

If we substitute $\psi_{\mathbf{k}}(\mathbf{r}) = u_{\mathbf{k}}(\mathbf{r})\, e^{i\,\mathbf{k}\cdot\mathbf{r}}$ into the Schrödinger equation with fixed **k**, where $u_{\mathbf{k}}(\mathbf{r})$ has the periodicity of the lattice, we find that $u_{\mathbf{k}}(\mathbf{r})$ is determined by the eigenvalue equation

$$H_{\mathbf{k}}\, u_{\mathbf{k}}(\mathbf{r}) = E_{\mathbf{k}}\, u_{\mathbf{k}}(\mathbf{r}) \tag{4.19}$$

which is related to the primitive cell, and where H is the Hamiltonian.

The solutions of (4.19) form a set of discrete eigenvalues $E_{n\mathbf{k}}$ for each **k**. In the limit of an infinite crystal the set of **k** values becomes dense in **k** space and we can write the energy eigenvalues as a continuous function of **k**, i.e. as $E_n(\mathbf{k})$. Note that since a real macroscopic crystal has of the order of 10^{23} atoms or molecules, it can be considered infinite in this context. The allowed energy values of an electron in a periodic potential can therefore be described as a set of continuous functions $E_n(\mathbf{k})$, which are known as *energy bands*.

It can be shown that not only one **k**, but all $\mathbf{k}' = \mathbf{k} + \mathbf{G}$, are associated with a $\psi_{n\mathbf{k}}(\mathbf{r})$. Since all points in **k**-space related by reciprocal lattice vectors **G** are equivalent, we can write $\psi_{n\mathbf{k}}(\mathbf{r}) = \psi_{n,\mathbf{k}+\mathbf{G}}(\mathbf{r})$, and $E_{n,\mathbf{k}+\mathbf{G}} = E_{n\mathbf{k}}$. Therefore, the representation of the energy bands $E_n(\mathbf{k})$ can be reduced to the (first) Brillouin zone. And since $E(\mathbf{k}) = E(-\mathbf{k})$, the bands can be represented for positive **k** only—the representation for both **k** and $-\mathbf{k}$ is redundant.

Further reading

Alcácer L 2018 From orbitals to bands *Electronic Structure of Organic Semiconductors* (Sao Rafael, CA: Morgan & Claypool) chapter 3 (https://doi.org/10.1088/2053-2571/aaddd8ch3)

Ashcroft N W and Mermin N D 1976 *Solid State Physics* (Philadelphia, PA: Saunders)

Kittel C 2004 *Introduction to Solid State Physics* 8th edn (New York: Wiley)

Problems

4.1 Consider a body-centred cubic crystal structure.
  (a) Calculate the structure factor.
  (b) Draw a cube and indicate the direction $[1\bar{1}1]$ and the plane $(1\bar{1}1)$.

4.2
  (a) Explain the role of x-ray diffuse scattering in the study of materials.
  (b) Figure 4.4 shows a $\mathbf{b}^*$ vector. What is its relation to the crystal structure? What are its dimensions?
  (c) The $\mathbf{q}_{SP}$ vectors (in red) are half of $\mathbf{b}^*$ and in the same direction; they end at different types of pattern (lines or spots) in the left and right images, and are not seen at room temperature. What can we infer has happened to the crystal structure?

4.3
  (a) Show that in a crystal lattice the translation operators commute with the electronic Hamiltonian.
  (b) Why is Bloch's theorem important for band theory?

References

[1] Close F 2000 *Lucifer's Legacy* (Oxford: Oxford University Press) p 77
[2] Bragg W H and Bragg W L 1913 *Proc. R. Soc. Lond.* A **88** 428–38
[3] Ravy S 2007 *Annu. Rep. Prog. Chem.* C **103** 223–60
[4] Pouget J P, Foury-Leylekian P and Almeida M 2017 *Magnetochemistry* **3** 13
[5] Bloch F 1928 *Z. Physik* **52** 555

Chapter 5

Physics of quasi one-dimensional conductors

Crystalline organic metals are highly anisotropic due to the nature of their crystal structures formed by stacks of molecules with 'π-systems'. They are realizations of quasi one-dimensional sets of interacting electrons. Like a line of cars in a narrow road, electrons in strict 1D systems cannot overtake each other. They can only move collectively, as described by the Luttinger liquid model, which will be briefly introduced. This overview of the physics of quasi one-dimensional conductors will be very concise; we refer the reader to the literature suggested in Further reading, in particular, the review article by Jérôme and Schulz, and chapters 12 and 27 in the book edited by Lebed.

5.1 Introduction

The theories that better describe the physics of crystalline organic metals and superconductors need to take into account the quasi one-dimensional character of the interacting many-electron system. The structural features provide electron delocalization and subsequent electron mobility predominantly along the stacking axis, leading to high anisotropy of transport and other properties. A simple band structure can be approximately calculated by the Hückel (tight binding) method based on the formation of energy bands due to the overlap between the π molecular orbitals of adjacent molecules. A linear combination of orbitals gives, in the case of one orbital per primitive cell corresponding to a regular stack of molecules,

$$E(k) = 2\, t_{\parallel} \cos(ka) \qquad \text{with} \quad k = 2\pi\, m/Na; \quad m = -N/2, \ldots, 0, \ldots, N/2 \qquad (5.1)$$

where a is the lattice spacing along the stack, N is the number of lattice sites, and $t_{\parallel} \equiv t_a$ is the transfer integral along the stack. The band width is $4t_{\parallel}$. When considering interchain interactions one must include the components of the energy dispersion and the respective transfer integrals between neighbours in the different directions.

In the Bechgaard salts, for example, the ratios between the bandwidths along the stacking axis and those along the perpendicular directions are of the order of $4t_a$: $4t_b$: $4t_c = 1\ eV$: $0.1\ eV$: $0.003\ eV$, and they are even larger in other cases. This is the reason for the emergence of the unusual phenomena expected for one-dimensional (1D) solids. Although this simple band model seems to be a reasonable starting point, interchain hopping and the relative importance of electron–phonon and electron–electron interactions must be taken into account.

Peierls [1] predicted that a one-dimensional electron system could not have metallic properties, and Fröhlich [2] found that the ground state of such a system interacting with the lattice displacements would be unstable when exposed to a lattice modulation with a wavelength that would open a gap in the energy dispersion at the Fermi level. According to Fröhlich, such periodic lattice displacements would lead to periodic fluctuations in the electron density, and the joint system of electrons and lattice displacements could move as a wave along the lattice without being disturbed. Electron excitations across the gap would lead to a decrease in the electron density fluctuations and the width of the gap.

The following instabilities can be found in one-dimensional interacting electron systems:

1. Peierls instability [1, 2], which may lead to the formation of charge or spin density waves [3, 4]; this generally competes with the superconducting Bardeen, Cooper, and Schrieffer (BCS) instability [5]. A charge density wave (CDW) always leads to a lattice distortion, while a spin density wave (SDW) may form without lattice modulation.
2. Thermal fluctuations can destroy the long-range order in one-dimensional electron systems with short-range interactions at finite temperatures ($T > 0$), [6].
3. Disorder in one-dimensional electron systems causes electron localization, leading to metal–insulator transitions [7].

We will now outline the physics of these instabilities and their main consequences, starting with an overview of the Peierls instability and the associated low-temperature CDW state, in the context of mean-field theory. We will then discuss the competition between the Peierls and the superconducting BCS instabilities, introducing the concept of the Luttinger liquid, a very powerful and elegant approach that describes the behaviour of one-dimensional interacting electron systems.

5.2 Mean field theory of the Peierls transition

The Peierls instability can develop in a one-dimensional electron system in a periodic lattice with lattice constant a and incommensurate or commensurate band filling. In a commensurate system, the Fermi wave vector, $k_F = \pi/2na$, where n (rational number, m/m') is the *order of commensurability*, (e.g. $n = 1$ for a half-filled band). The physics of such a system can be described in terms of electron–phonon interactions in several ways, namely using mean-field theory. Figure 5.1 illustrates the case of a half-filled band (i.e. $n = 1$), corresponding to one electron per site. The electron density is constant, $\rho(x) = \rho_0$, and the electron energy dispersion can be

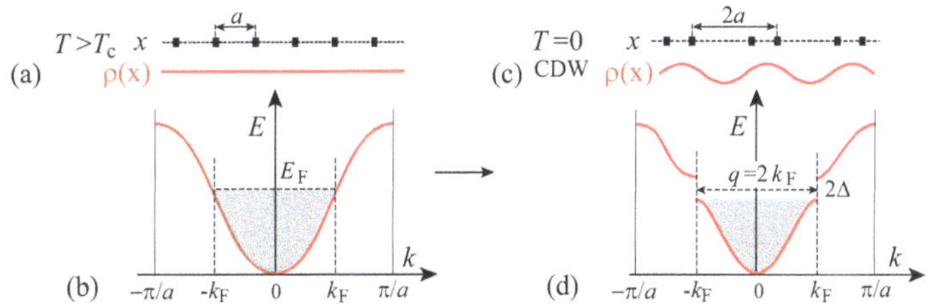

Figure 5.1. (a) One-dimensional lattice (with parameter a) and one electron per site. The electron density is constant, $\rho(x) = \rho_0$. (b) Energy dispersion of the undistorted one-dimensional electron system. (c) Distorted lattice and CDW, $\rho(x) = \rho_0 + \rho_1 \cos(2k_F\, x + \phi)$. (d) Energy dispersion after the Peierls transition. The energy of all states with $|k| \leqslant k_F$ has been decreased by the opening of a gap at k_F.

expressed by equation (5.1). Below a critical temperature T_c, the Peierls instability develops and leads to the formation of a CDW, $\rho(x) = \rho_0 + \rho_1 \cos(2k_F\, x + \phi)$, and the opening of a gap of 2Δ at a wave vector $q = 2k_F$ in the energy dispersion. At $T = 0$, the lattice distortion corresponds to its dimerization. The opening of the gap associated with the lattice distortion is favoured in one dimension, because in this case, the elastic energy required to distort the lattice is less than the gain in electron energy when the energy of all states with $|k| \leqslant k_F$ decreases. For bands that are more than half full, it is often convenient to think in terms of holes rather than electrons. For example, a band three-quarters full of electrons is equivalent to a band one-quarter full of holes. The Peierls instability can be described in the mean-field approximation by looking at the zero-temperature ground state using the Fröhlich Hamiltonian (in second quantization, see appendix A) to address the one-dimensional electron system interacting with a deformable lattice:

$$
\begin{aligned}
H &= H_{\text{el}} + H_{\text{ph}} + H_{\text{el-ph}} \\
&= \sum_{ks} \varepsilon_k c_{ks}^{\dagger} c_{ks} + \sum_{q} \omega_q b_q^{\dagger} b_q + \frac{g}{\sqrt{L}} \sum_{kqs} c_{k+q,\, s}^{\dagger} c_{ks} (b_q + b_{-q}^{\dagger})
\end{aligned}
$$

where H_{el}, H_{ph}, and $H_{\text{el-ph}}$ are the Hamiltonians for noninteracting electrons, for phonons, and for the electron–phonon interactions, respectively, and the wave vector summations are over the first Brillouin zone; $c_{ks}^{\dagger}$ and c_{ks} are creation and annihilation operators for electrons in Bloch state k and with spin s, with an energy ε_k relative to Fermi level; $b_q^{\dagger}$ and b_q are creation and annihilation operators for phonons with a wave vector q and a frequency ω_q; g is the electron–phonon coupling constant; and L is the length of the system. In the following we will use units such that $\hbar = 1$ and $k_B = 1$, and consider only electrons with wave vectors near k_F (i.e. $|k| \approx k_F$) and phonons with wave vectors near $2k_F$ (i.e. $|q| \approx 2k_F$), since those are the relevant ones in the Peierls instability.

In the phonon Hamiltonian (H_{ph}), the summation is over $q = 2k_F$ and $q = -2k_F$; we will use the mean values of $\langle b_{2k_F} \rangle = \langle b_{-2k_F}^{\dagger} \rangle$ and ignore the fluctuations around these values in this approximation. We can thus write

$$
H_{\text{ph}} = \sum_{q} \omega_q b_q^{\dagger} b_q = 2\omega_{2k_F}\, |\langle b_{2k_F} \rangle|^2 .
$$

To continue, we will define the order parameter $\Delta = |\Delta|\exp(i\phi) = 2g\langle b_{2k_F}\rangle/\sqrt{L}$, so that $\langle b_{2k_F}\rangle^2 = \Delta^2 L/4g^2$, and get the equation for the phonon Hamiltonian in terms of $|\Delta|$:

$$H_{\text{ph}} = \frac{L\omega_{2k_F}|\Delta|^2}{2g^2}.$$

We will now be concerned with the electronic Hamiltonian (the part involving the electrons and the electron–phonon interactions), using $(b_q + b_{-q}^{\dagger}) = 2\langle b_{2k_F}\rangle$ and the definition of the order parameter. We get:

$$\begin{aligned} H_{\text{el}}' &= H_{\text{el}} + H_{\text{el-ph}} \\ &= \sum_{ks} \varepsilon_k\, c_{ks}^{\dagger} c_{ks} + \Delta \sum_{k} c_{k-q,\,s}^{\dagger} c_{ks} + H.\,c. \end{aligned} \tag{5.2}$$

where H.c. stands for the Hermitian conjugate. To find the eigenvalues of H_{el}', we write (5.2) in matrix notation:

$$H_{\text{el}}' = \sum_{ks} \begin{pmatrix} c_{ks}^{\dagger} & c_{k-q,\,s}^{\dagger} \end{pmatrix} \begin{pmatrix} \varepsilon_k & \Delta^* \\ \Delta & \varepsilon_{k-q} \end{pmatrix} \begin{pmatrix} c_{ks} \\ c_{k-q,s} \end{pmatrix}$$

and diagonalize H_{el}', linearizing ε_k around $\pm k_F$ and get $\varepsilon_k = v_F(|k| - k_F)$, where v_F is the Fermi velocity. The secular determinant

$$\begin{vmatrix} v_F(|k| - k_F) - E & \Delta^* \\ \Delta & -v_F(|k| - k_F) - E \end{vmatrix} = 0$$

gives the solutions for the electronic energy E_k:

$$E_k = \text{sign}(|k| - k_F)[v_F^2(|k| - k_F)^2 + |\Delta|^2]^{1/2}.$$

We can now write the total Hamiltonian $H = H_{\text{el}} + H_{\text{ph}} + H_{\text{el-ph}}$ in the new basis as

$$H = \sum_{ks} E_k\, a_{ks}^{\dagger} a_{ks} + L\frac{\omega_{2k_F}}{2g^2}|\Delta|^2$$

where $a_{ks}^{\dagger}$ and a_{ks} are the creation and annihilation operators for electrons in the eigenstates of the total Hamiltonian. The total electronic energy can be obtained by summing over all states from $-k_F$ to $+k_F$ (see the Jérôme and Schulz article and Grüner's book in Further reading):

$$\begin{aligned} E_{\text{el}}(\Delta) &= 2\sum_{k=-k_F}^{k_F} E_k \\ &= -\frac{L}{\pi}\int_{-k_F}^{k_F} [v_F^2(|k| - k_F)^2 + |\Delta|^2]^{1/2} dk \\ &= -\frac{L\, n\, E_F}{2}\left[1 + \frac{|\Delta|^2}{E_F^2}\ln\left(\frac{2\, E_F}{|\Delta|}\right) + \text{higher order terms}\right] \end{aligned} \tag{5.3}$$

where $E_F = v_F k_F$ is the Fermi energy and n is the total electron density. The gain in electronic energy due to the opening of the gap given by equation (5.3), which is proportional to $-|\Delta|^2 \ln|\Delta|$, is always larger than the elastic energy loss of the lattice, which is only proportional to $|\Delta|^2$ (the increase in lattice energy is $(L\omega_{2k_F}/2g^2)|\Delta|^2$). We can therefore conclude that in the ground state ($T = 0$), the one-dimensional electron system is unstable; it tends to distort, ending up in a CDW state, with the opening of a gap in the electron energy dispersion (figure 5.1).

The value of the gap can be obtained by minimizing the total energy:

$$|\Delta| = 2\,E_F\,e^{-1/\lambda}, \qquad \lambda = \frac{2\,g^2}{\pi\,v_F\,\omega_{2k_F}}$$

where λ is the dimensionless electron–phonon coupling constant. Note that the gap opens at $\pm k_F$, implying that the wavelength of the lattice modulation is π/k_F.

The modulation of the lattice can be calculated by averaging the lattice displacement $u(x)$ in terms of the phonon operators:

$$u(x) = \sum_q \frac{1}{\sqrt{2L\omega_q}}(b_q + b_{-q}^\dagger)e^{iqx}.$$

This can be done by considering that the sum is taken over the two values $q = \pm 2k_F$, and that $\langle b_q + b_{-q}^\dagger \rangle = 2\langle b_{2k_F} \rangle = \sqrt{L}\,\Delta/g$, and $\Delta = |\Delta|\exp(i\phi)$, where ϕ is the phase of Δ.

$$\langle u(x) \rangle = \sqrt{\frac{2}{\omega_2 k_F}}\,\frac{|\Delta|}{g}\cos(2k_F\,x + \phi).$$

This modulation of the lattice is associated with the modulation of the electron density, and is known as a CDW.

The excited states are collective oscillations of both the lattice and the electron density, including the phase or amplitude of the CDWs. Fröhlich has shown [2] that a coupled CDW and lattice modulation can move, carrying a dc current through the crystal. Despite the existence of a gap in the energy dispersion, metallic conductivity would be expected ('Fröhlich conductivity'). Fröhlich argued that this mechanism could explain superconductivity. There are, however, mechanisms that may suppress the motion of CDWs, such as: (i) impurities (i.e. impurity pinning); (ii) 'locking' of the CDW into the lattice in commensurate systems in which the phase has a preferred value; (iii) in incommensurate partial charge-transfer salts that have CDWs in opposite directions. In addition to these collective excitations, single-particle excitations across the gap are possible, as in normal semiconductors.

The Peierls ordered state at $T = 0$ is expected to evolve as the temperature increases. The gap Δ decreases, and the long-range order completely disappears above a certain temperature. A remarkable result of the analysis of the lattice dynamical properties as a function of temperature shows that the effect of the electrons on the lattice implies that the onset of a CDW is associated with a sharp dip in the phonon frequency dispersion at the wave vector $q = k_F$, corresponding to a giant Kohn anomaly [8] and a distortion of the lattice, which can be observed by x-ray and neutron diffuse scattering (see figure 4.4), as precursor effects of the phase transitions. As the temperature is decreased, the dip in the phonon spectrum becomes deeper and deeper, reaching complete lattice softening at a critical temperature

$$T_c = 1.13\, E_F\, e^{-1/\lambda}$$

where λ is the electron–phonon coupling constant (note that $k_B = 1$ and $\hbar = 1$, as usual).

The description of the mean-field theory for the ground state of the Peierls instability given above shows many similarities with the BCS theory of superconductivity [9], namely: (i) the opening of a gap on the entire Fermi surface, implying a fast decrease in the electronic energy; (ii) the transition temperature and the zero-temperature gap only differ by the constant coefficient in the two cases. Both theories give $2|\Delta_{T=0}|/T_c = 3.52$, the well-known BCS relation between the zero-temperature gap and the transition temperature; (iii) the effect of backscattering impurities in the Peierls instability is identical to the effect of magnetic impurities in superconductors.

It should be added that in the Peierls state, electron–hole pairs are connected by the nesting relation $\varepsilon(k) = -\varepsilon(k - 2k_F)$, while in superconductivity, the Cooper pairs are two bound electrons with opposite momenta and spins. Curiously, both the Peierls electron–hole pairs and the Cooper pairs behave as bosons.

5.3 Rudiments of the Luttinger liquid model

The Luttinger liquid (LL) theory treats the low-energy excitations of a 1D interacting electron system as noninteracting bosons—actually as *sound waves*. In a 1D system the electrons cannot overtake each other. Therefore, permutation is not an issue, and they can be treated as bosons or fermions, the choice being a matter of convenience: strongly interacting fermions are equivalent to noninteracting bosons and vice versa. The concept was proposed by Tomonaga in 1950 [10], reformulated by Luttinger in 1963 [11], and improved by Mattis and Lieb [12]. The idea is linked to the fact that the Hamiltonian of the Tomonaga–Luttinger liquid is formally analogous to that of a vibrating string, and this fact can be used to highlight the collective character of the low-energy excitations of the 1D electron system.

The theory starts by linearizing the energy dispersion relation $\varepsilon(k)$ around $\pm k_F$ and extending the two branches '+' (also known as 'R') and '−' (also known as 'L'), representing electrons moving to the right ($k > 0$) and electrons moving to the left ($k < 0$), respectively, from $-\infty$ to $+\infty$ (figure 5.2). Since we are interested in low-energy excitations around the Fermi level, the introduction of the unphysical negative energies and the $-\infty$ to $+\infty$ domain does not affect the results, and simplifies the mathematics. The energy dispersion relation around $\pm k_F$ can then be written, taking E_F and k_F as reference values for ε and k, respectively (figure 5.2(a)):

$$\varepsilon_{R,L}(k) = \varepsilon_{\pm}(k) \approx \pm v_F(k - k_F)$$

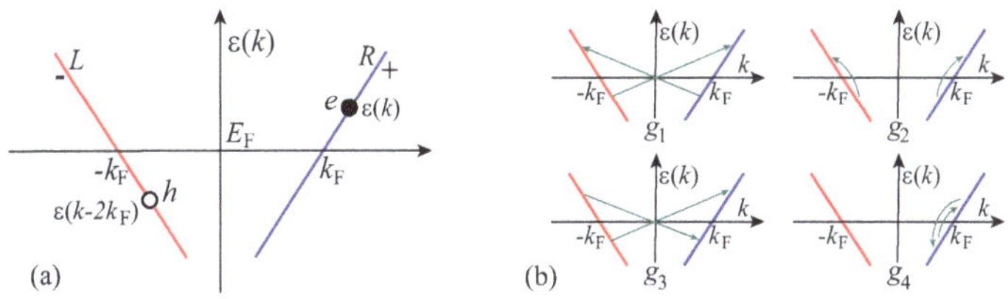

Figure 5.2. (a) Linearization of the energy dispersion around $\pm k_F$. Electrons and holes are connected through the nesting relation $\varepsilon(k) = -\varepsilon(k - 2k_F)$ and move with opposite velocities ($v_e = -v_h = v_F$). (b) Schematic representation of the scattering processes corresponding to the interaction coupling constants, $g_1 \ldots g_4$, in the so-called *g-ology* approach.

where $v_F(k - k_F)$ is obtained by expanding $\varepsilon(k)$ in a Taylor series around $\pm E_F$ and using $v_F = (1/\hbar)\, d\varepsilon/dk = d\varepsilon/dk$ (having set $\hbar = 1$). The linear dispersion implies a constant density of states, and that all excitations with momentum q have the same energy ε_q.

The electronic properties of the 1D electron system are determined by the interactions near the Fermi level, i.e. close to the Fermi surface, which in one dimension is reduced to the two points $\pm k_F$. The important low-energy processes due to the interactions are denoted by the coupling constants g_1, g_2, g_3, and g_4 (*the g-ology*). These processes are schematically illustrated in figure 5.2(b). In the g_1 process the fermions change sides of the Fermi surface in a backscattering event that involves a $2k_F$ momentum transfer. The g_2 is a forward-scattering process with a small momentum transfer, which couples fermions on opposite sides of the Fermi surface, but each fermion stays on its own side. The g_4 process is a similar forward-scattering process, but it only couples fermions on the same side of the Fermi surface. In the g_3 process a pair of fermions is scattered from $-k_F$ to $+k_F$, involving a total momentum transfer of $4k_F$, so that this process is only allowed if $4k_F$ is a reciprocal lattice vector, **G**; i.e. an Umklapp process, which may occur in a half-filled band system.

We now consider a model with spin 1/2 fermions and try to represent the excitations. The interactions are usually expressed in terms of the Fourier components of the particle density fluctuation operators for right- and left-moving fermions:

$$\rho_{\pm,s}(q) = \sum_k a^{\dagger}_{\pm,\, k+q,\, s} a_{\pm,k,s}$$

where $a_{k,s}$ annihilates a fermion in state k with spin $s = \uparrow, \downarrow$, and $a^{\dagger}_{k+q,\, s}$ creates a fermion in state $k + q$ with spin s, increasing the total momentum of the system by q. Here, $\rho_{\pm,s}(q)$ are density fluctuation operators with momentum q which contain two fermion operators, $a^{\dagger}$ and a that create a particle and a hole with opposite spins, and are therefore bosonic (they obey the boson commutation relations). Since, in a linearized energy dispersion, all electron–hole excitations with the same momentum q have the same energy, the density fluctuation operators satisfy the eigenvalue equation:

$$H\,\rho_{\pm,s}(q)\,|0\rangle = E_q\,\rho_{\pm,s}(q)\,|0\rangle.$$

H is the Hamiltonian and $|0\rangle$ is the ground state of the system. These are the only low-energy excitations possible in a 1D system and they are collective oscillations with momentum q.

The noninteracting Hamiltonian can be written in terms of fermion or boson operators:

$$\begin{aligned} H_0 &= v_F \sum_{k,s}\left[(k - k_F)a^{\dagger}_{+,\,k,\,s}a_{+,k,s} + (-k - k_F)a^{\dagger}_{-,\,k,\,s}a_{-,k,s}\right] \\ &= \frac{2\pi v_F}{L}\sum_{q>0,p,s}\rho_{p,s}(q)\rho_{p,s}(-q) \end{aligned}$$

where $p = \pm$ and L is the length of the system.

We can also introduce the interactions in terms of the density fluctuation operators ρ or in terms of the Fourier components which, for each branch $\pm$, are of the form:

$$V = \frac{1}{2}\int V(x - x')\rho(x)\,\rho(x')\mathrm{d}x\,\mathrm{d}x' \quad \text{or} \quad V = \frac{1}{2L}\sum_{k,\,k',\,q} V(q)\,a^{\dagger}_{k+q}\,a^{\dagger}_{k'-q}\,a_{k'}\,a_{k}. \tag{5.4}$$

The particle–particle (or more appropriately the density–density) interactions in (5.4), and illustrated in figure 5.2(b), are written in terms of the coupling constants g and the annihilation and creation operators (for example, the removal of a particle from states k, k' and creation in the states $k + q$, $k' - q$ as a result of scattering processes with total momentum exchange q).

The complete Hamiltonian of the 1D interacting electron system can then be written by adding the four interactions to the noninteracting Hamiltonian [13]:

$$\begin{aligned} H = v_F &\sum_{k,p,s}[(pk - k_F)\,a^{\dagger}_{p,\,k,\,s}a_{p,k,s}] \\ &+ \frac{1}{L}\sum_{k,q,s} g_1\,a^{\dagger}_{+,\,k_1+2k_F+q,\,s}\,a^{\dagger}_{-,\,k_2-2k_F-q,\,s'}\,a_{+,k_2,s'}\,a_{-,k_1,s} \\ &+ \frac{1}{L}\sum_{k,q,s} g_2\,a^{\dagger}_{+,\,k_1+q,\,s}\,a^{\dagger}_{-,\,k_2-q,\,s'}\,a_{-,k_2,s'}\,a_{+,k_1,s} \\ &+ \frac{1}{2L}\sum_{p,k,q,s} g_3\,a^{\dagger}_{p,\,k_1+p2k_F+q,\,s}\,a^{\dagger}_{p,\,k_2-p2k_F-q+pG,\,s'}\,a_{-p,k_2,s'}\,a_{-p,k_1,s} \\ &+ \frac{1}{L}\sum_{p,k,q,s} g_4\,a^{\dagger}_{p,\,k_1+q,\,s}\,a^{\dagger}_{p,\,k_2-q,\,-s}\,a_{p,k_2,-s}\,a_{-p,k_1,s} \end{aligned} \tag{5.5}$$

where $p = +$ refers to the right-moving fermions and $p = -$ refers to the left-moving fermions. We can check the terms of the interactions $g_1 \ldots g_4$ using the schematic representations of the scattering processes shown in figure 5.2(b).

To use the bosonization representation we introduce the field operators $\phi(x)$ and $\Pi(x)$ for the two spin values separately, and transform them into charge and spin boson operators, making $\phi_{\rho,\sigma} = (\phi_\uparrow \pm \phi_\downarrow)/\sqrt{2}$ and $\Pi_{\rho,\sigma} = (\Pi_\uparrow \pm \Pi_\downarrow)/\sqrt{2}$, which obey the boson commutation relations $[\phi_\nu(x), \Pi_{\mu'}(x')] = \mathrm{i}\delta(x - x')$. The Fermi field can be written as [14, 15]

$$\begin{aligned}\psi_{p,s}(x) &= \frac{1}{\sqrt{L}} \sum_k a_{p,k,s}\, \mathrm{e}^{ikx} \\ &= \lim_{\alpha\to 0} \frac{\mathrm{e}^{ipk_{\mathrm{F}}x}}{\sqrt{2\pi\alpha}} \exp\left\{-\frac{\mathrm{i}}{\sqrt{2}}[p(\phi_\rho + s\phi_\sigma) + (\theta_\rho + p\theta_\sigma)]\right\}\end{aligned}$$

where α is a short-distance cutoff, and $\theta_\nu(x)$ is related to $\Pi_\nu(x')$ by $\theta_\nu(x) = \pi \int \Pi_\nu(x')\mathrm{d}x'$.

In this representation, the full Hamiltonian for a 1D system of interacting electrons takes the form [13, 14, 16]:

$$H = H_\rho + H_\sigma + H_{g_1} + H_{g_3}.$$

In explicit form, this is:

$$\begin{aligned}H = \sum_{\nu=\rho,\sigma} \frac{1}{2\pi} &\int \left\{u_\nu K_\nu[\pi\, \Pi_\nu(x)]^2 + u_\nu K_\nu^{-1}[\nabla\phi_\nu(x)]^2\right\}\mathrm{d}x \\ &+ \frac{2g_1}{(2\pi\alpha_0)^2} \int \cos[\sqrt{8}\, \phi_\sigma(x)]\mathrm{d}x \\ &+ \frac{2g_3}{(2\pi\alpha_0)^2} \int \cos[\sqrt{8}\, \phi_\rho(x)]\mathrm{d}x\end{aligned} \tag{5.6}$$

where

$$u_\nu = \sqrt{\left(v_{\mathrm{F}} + \frac{g_{4,\nu}}{\pi}\right)^2 - \left(\frac{g_\nu}{2\pi}\right)^2}, \quad K_\nu = \sqrt{\frac{2\pi v_{\mathrm{F}} + 2g_{4,\nu} + g_\nu}{2\pi v_{\mathrm{F}} + 2g_{4,\nu} - g_\nu}} \tag{5.7}$$

and (since $\nu = \rho, \sigma$) $g_\rho = g_1 - 2g_2$, $g_\sigma = g_1$, $g_{4,\rho} = g_4$, and $g_{4,\sigma} = 0$. The last term in (5.6), the g_3 term, is only relevant if the Umklapp scattering process is to be taken into account.

From (5.7) we find that for a noninteracting system, $u_\nu = v_{\mathrm{F}}$, i.e. the charge and spin velocities are equal; and $K_\nu = 1$. If there is no backscattering ($g_1 = 0$), equation (5.6) describes independent charge and spin density oscillations and the system is conductive.

The two first terms, $H_\rho + H_\sigma$, i.e. the first line in equation (5.6), represent only the forward interactions (g_2) and (g_4), and correspond to the Tomonaga–Luttinger (TL) Hamiltonian, which is exactly solvable. This Hamiltonian is identical to the Hamiltonian of an elastic string, where u_ν are the velocities of the collective excitations and K_ν are the dimensionless stiffness constants of sound waves, which depend on the interactions. $\Pi_\nu(x)$ and $\phi_\nu(x)$ are Bose field operators corresponding to displacement and conjugate momenta. ϕ_ν are related to the long-wavelength

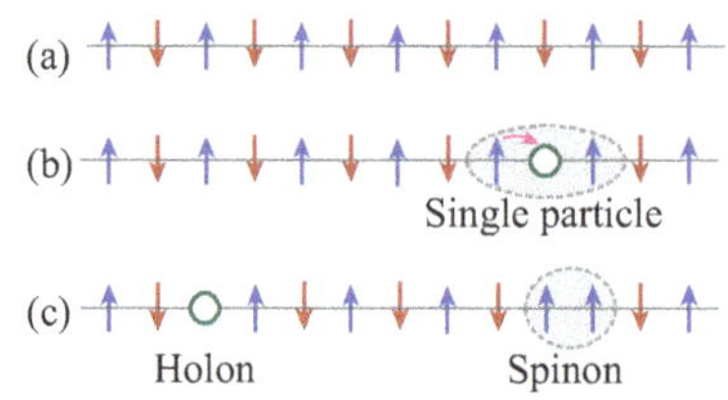

Figure 5.3. (a) Schematic representation of the ground state of a one-dimensional interacting electron system. (b) Removing one particle creates a single-particle excitation which is converted into two independent excitations containing separate charge and spin degrees of freedom: a *holon* and a *spinon*, respectively, that can propagate with different velocities (c).

distortions of the charge and spin densities, $\rho(x)$ and $\sigma(x)$, respectively, by $\rho(x) = -\sqrt{2}\,\nabla\phi_\rho(x)/\pi$ and $\sigma(x) = -\sqrt{2}\,\nabla\phi_\sigma(x)/\pi$. The energies are given by $\omega(q) = u|q|$ and the number current density is $j(x) = uK\Pi(x)$.

According to the TL Hamiltonian (the first line of (5.6)), there are no quasiparticles as in the Fermi liquid, but only two types of collective excitations (sound waves), one carrying charge and the other carrying spin, both governed by the two parameters, u_ν and K_ν. We can visualize a collective excitation of this kind using the following example. Imagine a line of cars in a one-lane road. If a car accelerates slightly in one direction, it will soon hit the next car and transmit its momentum to it; the next car will then hit a further one, and so on. The initial localized event will rapidly propagate along the whole system.

The remarkable fact that the Hamiltonian (5.6) separates into two independent parts, a charge part and a spin part, is illustrated in figure 5.3, where the removal of a fermion creates a kink in both ϕ_ρ and ϕ_σ, corresponding to a single-particle excitation which splits into two excitations that can propagate independently.

The exploration of the domains of the various processes depicted in figure 5.2(b) leads to the establishment of phase diagrams such as the one shown in figure 5.4, where the various competing instabilities, which refer to the same set of states, are described by the Hamiltonian (5.5). They are the CDW, the SDW, and the singlet superconducting (SS) and triplet superconducting (TS) phases. The phase domains are represented in the g_1–g_2 plane, the most stable being those with the highest transition temperatures. The $g_1 = 2g_2$ line separates the density wave phases ($g_2 > 0$) from the superconducting phases ($g_2 < 0$).

5.3.1 Power-law behaviour of the model

One of the most remarkable properties of the Luttinger liquid model is that all correlation functions follow power laws with exponents that may differ, but all depend on K_ρ. These power laws are usually associated with critical states, such as those occurring at critical temperatures in phase transitions. There is no step in the occupation number n_k (figure 2.4), as in a Fermi liquid, implying that quasiparticles do not exist in one dimension; the quasiparticle weight $Z(T)$ still has a singularity at k_F, but it decays with a power law: $Z(T) \sim T^\alpha$, where α is

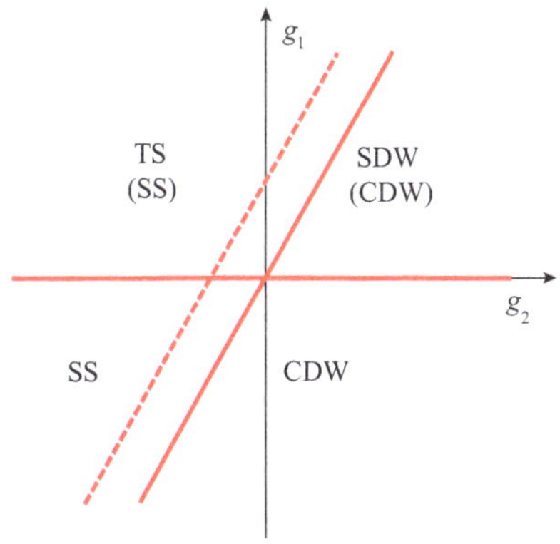

Figure 5.4. Phase diagram for the 1D interacting electron system described by the Hamiltonian (5.5) in the (g_1–g_2) plane, showing the domains for which the various phases are stable. In the presence of Umklapp scattering, the superconducting–antiferromagnetic (TS(SS)/SDW(CDW)) phase boundary is shifted to $g_1 = 2g_2 + |g_3|$ (dashed line) [17]. Phases in brackets have a lower degree of divergency in the response functions than the others.

$$\alpha = (K_\rho + 1/K_\rho - 2)/4$$

and

$$n(k) - n(k_F) \propto \text{sign}(k - k_F)|k - k_F|^\alpha .$$

α is also the exponent of the single-particle density of states at the Fermi level [13]:

$$N(0) \sim (\pi v_F)^{-1}\left(\frac{T}{E_F}\right)^\alpha .$$

The single-particle correlation function decays with time or distance with the exponent $\zeta = \alpha + 1$, and the $2k_F$ oscillating parts of the spin–spin and density–density correlation functions decay with an exponent $K_\rho + 1$ [16]. The power-law behaviour of the Tomonaga–Luttinger liquid has been probed by several experiments performed with highly anisotropic organic conductors; in particular, high-frequency conductivity measurements have been made, some of which are described below.

5.3.2 The effect of interchain coupling

The main effect of interchain coupling is the possibility of dimensional crossover between a one-dimensional situation and a higher-dimensional one. Let us consider the single-particle energy in the following case:

$$\varepsilon(k_\parallel, k_\perp) = -2\, t_\parallel \cos(k_\parallel a) - 2\, t_\perp \cos(k_\perp b)$$

where b is the perpendicular direction. In quasi-1D organic conductors $t_\parallel \gg t_\perp$, and the Fermi surface is open. Due to the interchain transfer integral $t_\perp$ the Fermi surface is warped, as shown in figure 5.5. If the temperature is larger than the warping of the Fermi surface due to interchain hopping (i.e. if $k_B T > w$), the system

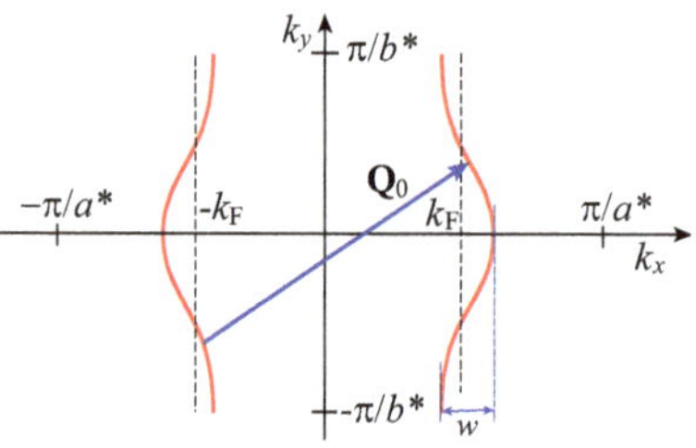

Figure 5.5. The Fermi surface of a quasi one-dimensional conductor showing the warping, w, due to the interchain transfer integral $t_\perp$. The nesting wave vector is $\mathbf{Q}_0 = (2k_F, \pi/b^*, \pi/c^*)$.

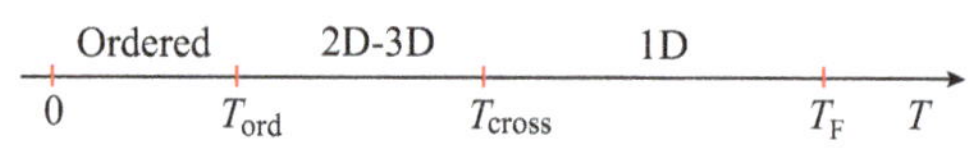

Figure 5.6. Temperature scales when $t_\perp \ll t_\parallel$. For temperatures higher than the intrachain hopping ($k_BT > t_\parallel$), or T_F, perturbation theory is valid. Below this temperature, the system is in an interacting 1D regime. At the crossover temperature T_{cross} the interchain hopping $t_\perp$ couples the chains and destroys the 1D character. For commensurate systems (Mott insulators) the gap can suppress the hopping and drive T_{cross} to zero. In all cases, the system can make a transition to an ordered state at a temperature T_{ord}. Adapted with permission from [18]. Copyright 2004 American Chemical Society.

is not impacted by the warping and behaves as a one-dimensional system. At lower temperatures, such that $k_BT < w$, the system is impacted by the warping and the 2D or 3D nature of the energy dispersion, and behaves as a 2D or 3D system—there is a dimensional crossover as the temperature is lowered (see figure 5.6). As long as the so-called ideal nesting condition $\varepsilon(\mathbf{k}) + \varepsilon(\mathbf{k} + \mathbf{Q}_0) = 0$ is preserved and electrons and holes move with opposite velocities $\mathbf{v}_e(\mathbf{k}) = -\mathbf{v}_h(\mathbf{k} + \mathbf{Q}_0)$, the electron–hole interactions are effectively one-dimensional, and in the case of attractive interactions they generate CDW or SDW semiconducting phases with the order parameter

$$\Delta_{CDW,SDW}(\mathbf{r}) = \Delta_{CDW,SDW}\cos(\mathbf{Q}_0\mathbf{r}).$$

In the presence of a magnetic field with $\mathbf{H} = (0, 0, H)$ and $\mathbf{A} = (0, Hx, 0)$, there is an instability of the metallic phase with respect to the formation of field-induced spin density waves (FISDWs), and, in consequence, electrons and holes can only move along x, i.e. along the conducting chain. This means that the magnetic field decreases the dimensionality of the system. In principle, an infinite series of FISDW phase transitions to different quantum Hall effect states is possible, although a limited number can be observed in practice [19, 20].

For commensurate systems, such as most organic conductors, the situation is even more complicated than for the incommensurate systems. In the commensurate systems one expects to observe Mott insulator (MI) behaviour at low temperatures. Near absolute zero, thermal fluctuations are reduced, and quantum fluctuations become predominant. When quantum fluctuations resulting from competing ground

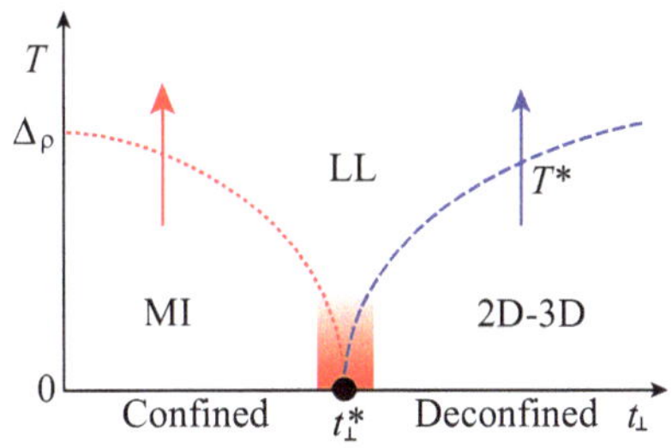

Figure 5.7. Schematic phase diagram for a correlated electron system with an ordered state next to a metallic state. At $T = 0$, a *deconfinement* transition occurs for $t_\perp = t_\perp^*$ at which the MI state is destroyed and the system becomes a 2D–3D metal. Additional complications may occur near the quantum critical point represented by the shaded rectangle. If $t_\perp$ is small the system is a magnetic insulator with a gap Δ_ρ. As the temperature is increased above the gap ($k_B T > \Delta_\rho$) a crossover to a 1D LL occurs (dotted line). For materials with $t_\perp > t_\perp^*$, a 2D or 3D metal is expected at low T; at $T = T^*$ a crossover occurs (dashed line) from a 2D–3D FL to an LL. The crossovers are indicated by the arrows. Adapted with permission from [18]. Copyright 2004 American Chemical Society.

states are strong enough, one may have a second-order phase transition at $T = 0$, i.e. a *quantum phase transition*. In the vicinity of the *quantum critical point* (QCP), anomalous magnetic and metallic behaviours, such as unconventional superconductivity, may be observed. In organic conductors one may consider a *deconfinement quantum transition* (at $T = 0$), in which electrons cease to be confined to their own chains. Interchain hopping can then destroy the magnetic ordered insulating (MI) state in the range of the quantum critical point $t_\perp^*$, above which the system becomes a 2D or 3D Fermi liquid (FL). This situation is illustrated using a phase diagram as a function of the temperature, T, and the interchain transfer integral, $t_\perp$, in figure 5.7. The metal–insulator transition and the Luttinger liquid physics have drastic consequences for the transport properties of systems of weakly coupled chains. These properties can be used as a probe for the Luttinger liquid behaviour. If one ignores the renormalization of K_ρ by the Umklapp process, for frequencies larger than those corresponding to the Mott gap, Δ_ρ, [16] one gets

$$\sigma_\parallel(\omega) \propto \omega^{4n^2K_\rho-5}, \qquad \sigma_\perp(\omega) \propto \omega^{2\alpha-1} \tag{5.8}$$

for the a.c. longitudinal and transverse conductivities, respectively, where n is the order of commensurability which is related to the band filling, $1/2n$. A similar power-law behaviour can be seen in the temperature-dependent transport for the resistivities along and perpendicular to the chain axis:

$$\rho_\parallel(T) \propto T^{4n^2K_\rho-3}, \qquad \rho_\perp(T) \propto T^{1-2\alpha}.$$

For a one-quarter full band system, $K_\rho < 0.25$.

The Luttinger liquid model gives a remarkable insight into the transport properties of quasi one-dimensional organic conductors. The optical conductivity is an important probe of the Luttinger liquid behaviour, and one can get the K_ρ parameter experimentally.

5.4 Conclusions

In this chapter we have very briefly presented the main concepts relevant to the physics of quasi one-dimensional conductors, from the Peierls and spin-Peierls instabilities, CDWs, and SDWs to the more complex phenomena of superconductivity and crossover and deconfinement transitions between the one-dimensional insulating state and a high-dimensional metal. The rich and fascinating physics found in these systems stems from the interactions between the one-dimensional chains which characterize the crystal structures of organic conductors and superconductors.

Further reading

Giamarchi T 2004 *Quantum Physics in One Dimension* (Oxford: Clarendon)
Grüner G 1994 *Density Waves in Solids.* (Cambridge, MA: Perseus)
Schulz H, Cuniberti G, and Pieri P 1998 *Fermi liquids and Luttinger liquids* (https://arxiv.org/abs/cond-mat/9807366)
Jerome D and Schulz H J 2002 *Adv. Phys.* 51 293–479 (https://doi.org/10.1080/00018730110116362)
Lebed A G (ed) 2008 *The Physics of Organic Superconductors and Conductors* (Berlin: Springer)

Problems

5.1 Describe some of the main differences and similarities of the Peierls transition and the BCS type of superconductivity.
5.2 Explain why a one-dimensional lattice with a three-quarters full electron band may lead to a tetramerization of the lattice.
5.3 Explain why electrons in a strictly 1D system can be considered as bosons and as fermions.

References

[1] Peierls R E 1955 *Quantum Theory of Solids* (Oxford: Oxford University Press)
[2] Fröhlich H 1954 *Proc. R. Soc.* A **223** 296–305
[3] Slater J C 1951 *Phys. Rev.* **82** 538
[4] Overhauser A W 1960 *Phys. Rev. Lett.* **4** 462–5
[5] Bychkov Y A, Gor'kov L P and Dzyaloshinskii I E 1966 *Sov. Phys. JETP* **23** 489–501
[6] Hohenberg P C 1967 *Phys. Rev.* **158** 383–6
[7] Mott N F and Twose W D 1961 *Adv. Phys.* **10** 107–63
[8] Kohn W 1959 *Phys. Rev. Lett.* **2** 393
[9] Bardeen J, Cooper L N and Schrieffer J R 1957 *Phys. Rev.* **108** 1175–204
[10] Tomonaga S 1950 *Prog. Theor. Phys.* **5** 544–69
[11] Luttinger J M 1963 *J. Math. Phys.* **4** 1154–62
[12] Mattis D C and Lieb E H 1965 *J. Math. Phys.* **6** 304
[13] Bourbonnais C and Jerome D 2008 Interacting electrons in quasi-one-dimensional organic superconductors *The Physics of Organic Superconductors and Conductors* ed A G Lebed (Berlin, Heidelberg: Springer) pp 357–412

[14] Schulz H, Cuniberti G and Pieri P 1998 (arXiv:cond-mat/9807366v2) [cond-mat.str-el] (https://arxiv.org/abs/cond-mat/9807366)
[15] Giamarchi T 2004 *Quantum Physics in One Dimension* (Oxford: Clarendon)
[16] Giamarchi T 2008 From Luttinger to Fermi Liquids in Organic Conductors *The Physics of Organic Superconductors and Conductors* ed A Lebed (Berlin: Springer) pp 719–43 ch 27
[17] Jérome D, Creuzet F and Bourbonnais C 1989 *Phys. Scr.* **T27** 130–5
[18] Giamarchi T 2004 *Chem. Rev.* **104** 5037–56
[19] Gor'kov L and Lebed A 1984 *J. Phys. Lett.* **45** 433–40
[20] Lebed A G (ed) 2008 *The Physics of Organic Superconductors and Conductors* (Berlin, Heidelberg: Springer)

Chapter 6

Organic conductors and superconductors

This chapter focuses on two representative case studies of organic conductors and superconductors. The first is $(TMTSF)_2PF_6$, the first organic superconductor to be synthesized, in 1979, and a prototype example for low-dimensional physics. The other case study is the perylene–metal dithiolate series, of which the first members were synthesized in the late 1960s. They are highly one-dimensional and come in magnetic and nonmagnetic flavours. Both cases are good examples of close collaboration between chemists and physicists, and strong interaction between experiment and theory.

6.1 The wondrous $(TMTSF)_2PF_6$

Di-(tetramethyltetraselenafulvalene)-hexafluorophosphate, $(TMTSF)_2PF_6$, was the first organic superconductor to be reported. It behaves as a metal with a resistivity of about 1 $\mu\Omega$ cm at 4 K, and becomes a superconductor under pressure with a transition temperature of 0.9 K [1, 2]. It is a nonconventional superconductor and the first system to exhibit the three-dimensional quantum Hall effect. $(TMTSF)_2PF_6$ is one of the Bechgaard salts, a series of highly conducting salts based on the TMTSF molecule, namely $(TMTSF)_2X$ where X = PF_6^-, AsF_6^-, SbF_6^-, BF_4^- and NO_3^-, which were synthesised and studied [1] as a result of the search for (high-temperature) organic superconductors and the collaboration between chemists and physicists, in this case between Klaus Bechgaard and Denis Jérome.

6.1.1 Molecular and crystal structure

The molecular structure and side and top views of the highest occupied molecular orbital (HOMO) of TMTSF are shown in figure 6.1. A detailed analysis of energy levels, based on near-edge x-ray absorption fine-structure measurements and density functional theory (DFT) calculations, was performed by Medjanik *et al* [3] for both TMTSF and PF_6. The value of the HOMO–lowest unoccupied molecular orbital

doi:10.1088/978-0-7503-3347-4ch6

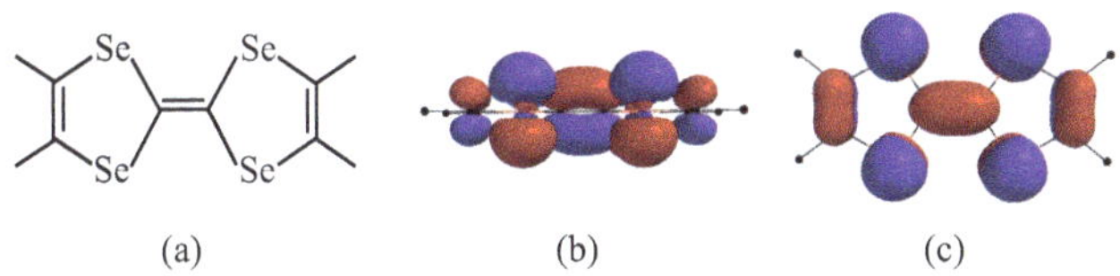

Figure 6.1. TMTSF molecule (a) and side (b) and top (c) views of the highest occupied molecular orbital (HOMO), as obtained from a DFT/B3YLP calculation.

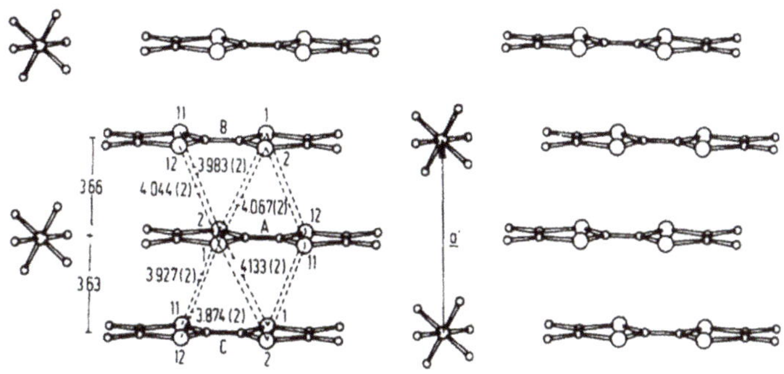

Figure 6.2. Crystal structure of $(TMTSF)_2PF_6$. Side view of stacks (tilted by 10°); a' is the projection of **a** [4]. Reproduced with permission from the International Union of Crystallography, according to the terms and conditions of use of material published by the International Union of Crystallography.

(LUMO) gap of the TMTSF molecule obtained from DFT calculations based on the geometry of TMTSF in the crystal structure of $(TMTSF)_2PF_6$ at 4 K was 1.57 eV. The crystal structure of $(TMTSF)_2PF_6$ as originally determined by Thorup *et al* [4] at ambient pressure and temperature (295 K) is shown in figure 6.2. It is triclinic $P\bar{1}$ with $a = 7.297$ Å, $b = 7.711$ Å, $c = 13.522$ Å, $\alpha = 83.39°$, $\beta = 86.27°$, and $\gamma = 71.01°$.

The nearly planar TMTSF molecules are almost perpendicular to **a** (with a tilt of 1.1°) and form stacks along **a**. Within each stack the TMTSF molecules repeat by an inversion, leading to zigzag stacking and two slightly different interplanar distances (3.63 and 3.66 Å). This structure corresponds to a slight dimerization and a consequent split of the conduction band formed by the overlap of the HOMOs of the TMTSF molecules into a filled lower sub-band separated from a half-filled upper band by a gap, called the *dimerization gap*. Due to the finite transverse dispersion, the dimerization gap does not lead to a real gap in the density of states, according to extended-Hückel band calculations [5]. As a result, $(TMTSF)_2PF_6$ behaves as a system with a three-quarters filled band of electrons (one-quarter filled with holes). Note that the unit cell has two TMTSF molecules with a single + charge.

The bandwidths have been estimated from the thermopower and the plasma frequencies to be $4t_a = 1$ eV, $4t_b = 0.1$ eV, and $4t_c = 0.003$ eV [1, 6]. From an optical point of view, $(TMTSF)_2PF_6$ is considered to be a 1D metal at high temperatures, and a 2D or 3D system at low temperatures. The crossover from one regime to the other is expected to occur when $k_BT \sim 4t_\perp$, where $4t_\perp$ is the transverse bandwidth ($t_\perp = t_b$). However, the one-electron band model overestimates the value for the dimensional crossover because the electrons interact.

6.1.2 Electronic properties

The room-temperature (T = 300 K) dc conductivities of $(TMTSF)_2PF_6$ were measured in single crystals along the three crystallographic directions. The following values were found: $\sigma_{\|} = 540\ \mathrm{S\ cm^{-1}}$, $\sigma_{\perp 1} = 1.5\ \mathrm{S\ cm^{-1}}$, and $\sigma_{\perp 2} = 0.015\ \mathrm{S\ cm^{-1}}$ [1]. $\sigma_{\|}$ decreases monotonically from room temperature to a minimum at 18 K. From the maximum slope of the curve, it has been inferred that a metal–insulator transition occurs at 15 K (figure 6.3(a)).

The nature of this transition and the ground state of the system were subsequently established to be a spin density wave (SDW) by nuclear magnetic resonance (NMR) experiments, namely the ^{77}Se NMR line, confirming the metal–insulator transition temperature at 12 K [7]. Later diffraction experiments provided evidence that the insulating ground state is, in fact, a mixture of a $2k_F$ SDW and a $2k_F$ charge density wave (CDW) [8, 9], indicating that a gap opens in the electronic spectrum. The application of pressure reduces the metal–insulator transition temperature, and at pressures near 10 kbar the transition vanishes and a superconducting (SC) transition takes place at $T_c = 0.9$ K [2]. The magnetization curves are those of a type-II

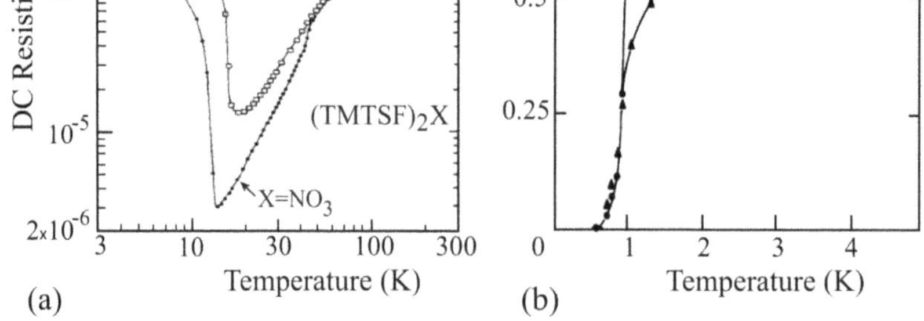

Figure 6.3. (a) Temperature dependence of the resistivity in typical samples of $(TMTSF)_2X$, including $(TMTSF)_2PF_6$. Note the logarithmic temperature and resistivity scales. Reprinted from [1] with permission from Elsevier. (b) Temperature dependence of the resistance normalized to the 4.6 K value for two crystals of $(TMTSF)_2PF_6$ at 12 kbar, showing the metal–superconductor transition at 0.9 K. Reprinted from [2] with permission from EDP Sciences.

superconductor [10]. Figure 6.3(b) illustrates the metal–superconductor transition of two crystals of $(TMTSF)_2PF_6$ under a pressure of 12 kbar.

The Seebeck coefficient, or thermopower, S, which is a measure of the entropy per carrier, is of the order of 20 $\mu V\,K^{-1}$ in the metallic regime; it falls linearly with temperature and can be extrapolated to zero at 0 K. For a highly anisotropic free electron gas, a three-quarters-full band contribution to S [1, 11] is given by

$$S = \frac{2\sqrt{2}}{3}\pi^2 \frac{k_B}{e}\frac{k_B T}{W}. \tag{6.1}$$

A bandwidth of $W = 1.0$ eV is found, which is in agreement with the optical results. Below the metal–insulator transition temperature the thermopower is indicative of semiconducting behaviour [1]. A recent study of the Seebeck coefficient and resistivity by Machida *et al* [12] revealed that below 1 K the resistivity exhibits a variable-range hopping temperature dependence and a Seebeck coefficient which increases as the temperature decreases and reaches a magnitude as large as 37 $mV\,K^{-1}$ at $T \sim 0.1$ K. This value, of the order of 400 k_B/e, is due to residual carriers hopping from one trap to another. According to the authors, these results give the first real evidence for the prediction that hopping electrons in the presence of Coulomb interaction can lead to a large Seebeck coefficient in the zero temperature limit. It should be mentioned that a giant Nernst signal (thermoelectric effect under a magnetic field) was observed in 2004 for magnetic fields aligned with the crystallographic axes in $(TMTSF)_2PF_6$ [13].

The magnetic susceptibility, measured using a Faraday balance, clearly shows a phase transition at $\approx$11.5 K, but the electron-spin-resonance signal disappears at 16–17 K [14].

6.1.3 The origin and nature of phase transitions

Pure 1D systems are better described by the Tomonaga–Luttinger liquid model. Like the Fermi liquid, which fails in 1D, this model only accounts for electron–electron interactions and ignores all other excitations, including phonons and their interactions with electrons.

In the pure 1D systems of interest, the relevant electron–electron interactions have momentum transfers of $2k_F$, zero, and $4k_F$, and are described in terms of the g_1, g_2 and g_3 coupling constants, respectively. The $4k_F$ scattering involves the transfer of two electrons from one side of the flat Fermi surface to the other, from which it is separated by $2k_F$. This process is allowed only if $4k_F$ is a reciprocal lattice vector (Umklapp scattering for a half- or quarter-filled band).

In quasi-1D conductors the transverse transfer integral $t_\perp$ is not really zero and the Fermi surface is warped. Therefore, there is a crossover temperature, T_{cross}, related to $t_\perp$, below which the system is impacted by the warping of the Fermi surface (Figure 5.5) and the interchain hopping $t_\perp$ couples the chains and destroys the 1D character (figure 5.6). The value of T_{cross} has been a matter of controversy for some time, but the most recent optical and transport measurements [15–18] place its value at about 100 K.

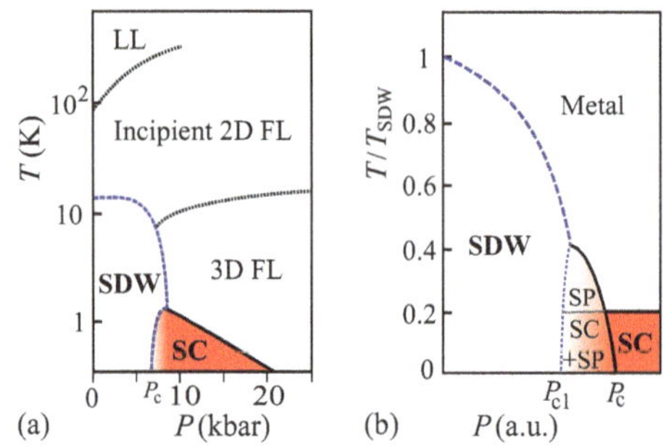

Figure 6.4. (a) Schematic temperature–pressure phase diagram for $(TMTSF)_2PF_6$, showing the Luttinger liquid (LL), the incipient 2D and 3D Fermi liquids (FLs), and the low-temperature insulating spin density wave (SDW) and superconducting (SC) phases. After [15], copyright Springer, 1998, reprinted by permission from Springer. (b) Detail of the phase diagram near the quantum critical point. SP stands for *soliton phase*. The continuous line corresponds to a first-order phase transition. Reprinted from [20] with permission from IOP.

The application of pressure increases the transfer integrals, and in particular the interchain transfer integral $t_\perp$, increasing the dimensionality of the system. This reduces the temperature of the metal–insulator transition and eventually suppresses it and induces an SC ground state. A pressure–temperature phase diagram of $(TMTSF)_2PF_6$ is schematically shown in figure 6.4(a); it indicates the metallic phase, the SDW, and the SC phases. For more details of the phase diagram see, for example, [19]. Figure 6.4(b) focuses on the short pressure range in the vicinity of the quantum critical point.

The critical pressure $P_c \approx 6.5$ kbar indicates the beginning of the SC low-temperature state, but in a rather fuzzy and unusual way. In fact, Greene and Engler [21] measured an increase of the resistivity between 4 and 1.2 K at $P = 6.5$ kbar, and found $P_c \approx 6.5$ kbar. The same value was found by Andres *et al* [10], but Jerome *et al* [2] found $P_c \approx 9$ kbar. This region between 6.5 and ≈9 kbar at a very low temperature, near the quantum critical point (see figure 5.7), has always been a centre of attention. In fact this phase diagram is typical of correlated electron systems with an ordered state next to a metallic state in which the tuning parameter is the pressure (acting on $t_\perp$). Although the metallic phase can usually be described as a Fermi liquid, signs of non-Fermi liquid behaviour appear above the quantum critical point [22]. Unconventional superconductivity is often found in this region of the phase diagram. The region around P_c has been thoroughly investigated, and it is clear that there is a pressure domain in which the competing SDW and superconductivity regimes coexist (see, for example, [21, 23–25]), although its real nature is still not settled.

It is well known that in a type-II superconductor, the region between the critical fields H_{c1} and H_{c2} is in a mixed or vortex state characterized by a distribution of the normal state within the SC material. It is also known that for a thin SC slab, the upper critical field in a parallel field configuration is greatly enhanced [26]. As the field is increased, a domain structure develops, characterized by a series of thin SC slabs sandwiched between SDW insulating phases. In a model suggested by Lee *et al* [27], the slabs arrange themselves so as to be perpendicular to the field direction. According to these authors, these slabs would correspond to the 'fuzzy' pressure domain in which the SDW and SC regime coexist, i.e. approximately between P_{c1} and P_c in figure 6.4(b). Further results reported by Lee *et al* [28] of simultaneous NMR and electrical transport experiments in this pressure range indicate a *tricritical point* (seen in figure 6.4(b)) that separates the line of the second-order SDW/metal transition (dashed line) from the line of first-order SDW/SC transitions along with the coexistence of the SDW and SC domains.

However, the real situation appears to be more complex, as shown, for example, by Kang *et al* [29] and Pasquier *et al* [30], who experimentally demonstrated the existence, in the pressure interval $P_{c1} < P < P_c$, of 1D and 2D SC domains with a crossover from bubbles to a filamentary superconductivity that first emerges along the c^* direction, towards slabs perpendicular to the a-axis, which merge in a homogeneous phase at P_c. According to these authors the only theory that seems to explain this domain formation is a *soliton model*, as proposed by Brazovskii *et al* [31] and by Gor'kov *et al* [20, 32], who suggested that in the pressure region $P_{c1} < P < P_c$ (in the vicinity of the quantum critical point) a new spatially inhomogeneous phase is formed by soliton domain walls (metallic/SC) perpendicular to the a-axis as a result of the change in the transfer integrals induced by the applied pressure. This would be a *soliton phase* (SP) that gradually develops at $P > P_{c1}$. In this phase, the SDW is gradually destroyed by pressure above P_{c1}, and the SC domains are expected to have a triplet character.

6.1.3.1 The superconducting state

The nature of the SC state in $(TMTSF)_2PF_6$, i.e. whether the spins of the Cooper pair are in a singlet state (SS) or in a triplet state (TS), became a matter of controversy soon after the discovery of organic superconductivity. Jérome, in his review paper of 1982 [33], and later in 1985 [34], had already suggested that superconductivity in these organic compounds is of the triplet rather than the singlet type, based on the evidence for repulsive electron–electron interactions, which, in the mean-field approximation, can lead to stable triplet superconductivity. He argued that the existence of a competing SDW, but not a CDW, implies that the coupling constant $g_1 > 0$ and that, as predicted in the g-ology phase diagram shown in figure 5.4, the most probable SC state is of the triplet type. The controversy was 'allegedly' settled in 2001 by Lee *et al* [35], who measured the Knight shift K_s to probe the spin susceptibility χ_s, and found that it remained unchanged through the SC transition, indicating that the SC state was triplet superconductivity. In a beautiful set of experiments Lee *et al* simultaneously measured the ^{77}Se NMR spectra, the spin–lattice relaxation rate, and the resistivity, vs temperature. The results were clear. It is a well-known fact that the Knight shift is proportional to the spin susceptibility, and that the spin susceptibility is proportional to the density of normal electrons, which vanishes at $T = 0$ for a singlet superconductor, but not for a triplet superconductor. In these experiments the Knight shift remained unchanged on cooling into the SC state, indicating that $(TMTSF)_2PF_6$ behaved as a triplet-type superconductor, at least under the two different sets of conditions in which the experiments were performed: $H = 2.38$ T, $p = 6.4$ kbar and $H = 1.43$ T, $p = 7$ kbar.

However, in 2007 Shinagawa *et al* found that for $(TMTSF)_2ClO_4$, the Knight shift did change when the SC transition temperature was passed, indicating singlet-type superconductivity, and there was no reason for $(TMTSF)_2PF_6$ to have a different type of superconductivity. This inconsistency with Lee *et al*'s results can be explained by the fact that Lee *et al*'s experiments with $(TMTSF)_2PF_6$ were conducted under a relatively high magnetic field (1.43 T). In fact there was evidence in the NMR relaxation rate vs magnetic field data of a crossover, or even a phase transition, occurring at low temperature and under a magnetic field of between 1 and 2 T, from a low-field singlet SC phase to a high-field SC triplet state phase or an inhomogeneous Fulde–Ferrell–Larkin–Ovchinnikov (FFLO) state [36]. The FFLO state is a high-field SC state which occurs in quasi-1D systems in which a Cooper pair forms from $\mathbf{k}_{F\uparrow}$ and $-\mathbf{k}_{F\downarrow}$ electrons with a nonzero centre of mass momentum $\mathbf{q}_{FFLO} = \mathbf{k}_{F\uparrow} - \mathbf{k}_{F\downarrow}$ and develops spatial oscillations in the order parameter. The competition between the singlet and the triplet regimes and the dimensional crossover (3D/1D) between the two have recently been discussed in [37, 38]. It should also be added that the triplet state (not the singlet) is a known instability of 1D or quasi-1D models [39, 40].

6.1.4 Field-induced spin density waves and the quantum Hall effect

$(TMTSF)_2PF_6$ is also notable for being the first material that exhibited the quantum Hall effect in a bulk crystal. It also exhibits cascades of field-induced spin density wave (FISDW) transitions to the quantum Hall-effect (QHE) states [41–46] (see section 2.4).

These magnetic phenomena justify the addition of a magnetic field coordinate to the pressure–temperature (P–T) phase diagram. As seen above, pressure increases the dimensionality of the system, and we now show that a magnetic field decreases it. The FISDW phases appear as a result of one-dimensionalization of the quasi-1D electron spectra due to the orbital electron effect in a magnetic field. In a magnetic field the dispersion consists of a set of 1D bands in a Brillouin zone defined by a *magnetic reciprocal lattice vector* $G = eHb/\hbar$ (b is the b-axis and H is the magnetic field) and the nesting can occur at $2k_F \pm nG$, where n is an integer. The cascade of FISDW transitions to the QHE states $i = 0, 1, 2, \ldots$ which emerges from the competition between the SDW $2k_F$ wave vector and the magnetic reciprocal lattice vector nG occurs as the quantum number n takes on high values for small fields and decreases to the conventional SDW with $n = 0$ for high fields. The values of n which label the FISDW subphases are the values of i which label the plateaux of the transverse resistivity $\rho_{xy} = h/i\, e^2$ (equation 2.15 and figure 2.3), as proven by Poilblanc *et al* [47]. However, according to Héritier [48], this mechanism for the quantization of the Hall effect results from cooperative phenomena that set the FISDW wave vectors so that the Fermi levels stay in the gaps of the itinerant states, this being therefore a different mechanism from the QHE observed in the 2D electron gas of semiconductor heterojunctions.

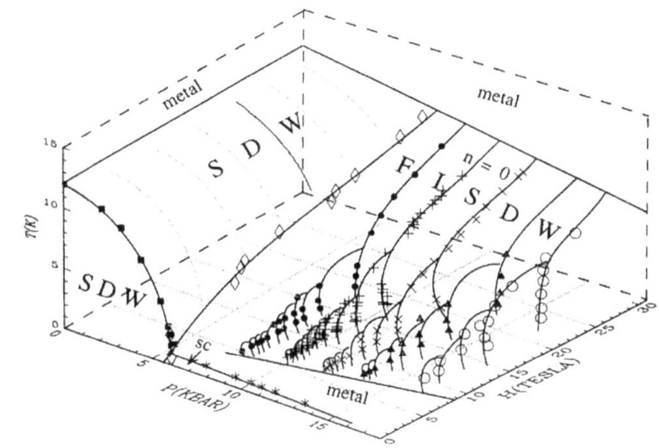

Figure 6.5. Pressure–magnetic-field–temperature phase diagram of $(TMTSF)_2PF_6$. The lines are guides for the eye. Copyright 1993 by the American Physical Society, reprinted from [43] with permission.

To make a long story short, it is instructive to look at the *P–H–T* phase diagram, reproduced in figure 6.5, where we can see the most remarkable aspects of $(TMTSF)_2PF_6$ as *P*, *H*, and *T* change: metallic, SDW insulating, SC, and FISDW–QHE phases.

As shown in the phase diagram, $(TMTSF)_2PF_6$ is a metal at room temperature, and at ambient pressure undergoes a transition to a SDW insulator at 12 K. At low temperature and ≈6 kbar, the SDW is suppressed, and superconductivity is observed at ≈1 K. The SC state is suppressed by a magnetic field of ≈0.2 T perpendicular to the conducting axis, and at ≈3 T a cascade of FISDW transitions to different QHE states emerges. The SDW transition temperature decreases with increasing pressure, and superconductivity appears at ≈6 kbar. At this pressure the SDW transition occurs prior to the SC transition, and the SDW insulating phase reappears as the magnetic field reaches the SC critical field. When the pressure is further increased up to ≈8 kbar, the first series of eleven distinguishable FISDW transitions emerges at a threshold field of more than about 4.5 T, and develops until it reaches the insulating, $n = 0$, FISDW state. New series of transitions appear as the pressure is increased, with an increase of the threshold field from 4.5 T at 8 kbar to 8 T at 16 kbar. The threshold field for the $n = 0$ transitions also increases from about 16 to 24 T between 8 and 16 kbar (figure 6.5).

This *P–H–T* phase diagram has been explained in terms of changing the dimensionality of the system [49–52]. The general shape depends on the unnested interchain transfer integrals along the *b* and *c* directions: pressure increases the dimensionality of the system by increasing the interchain transfer integrals; the magnetic field applied perpendicular to the conduction direction favours the 1D motion of the electrons. Note that the details of the phase diagram near the SDW–SC transition are not shown, but were discussed above.

Well-defined QHE plateaux corresponding to the integers $n = 5$ to 1 have been observed in single crystals of $(TMTSF)_2PF_6$ under pressure and at temperatures

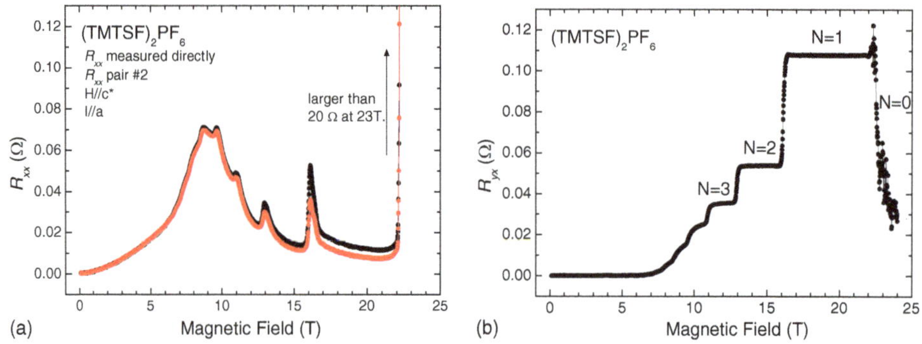

Figure 6.6. Magnetoresistances of $(TMTSF)_2PF_6$: longitudinal, R_{xx}, (a); and transverse, R_{yx}, (b), measured at 0.5 K and less than 10 kbar. Copyright (2004), reprinted from [53] with permission from Elsevier.

down to 0.5 K in magnetic fields of up to 25 T applied along the c^* direction [41, 42]. The quantization of the Hall effect is a consequence of the quantized nesting model mentioned above, which is used to interpret the FISDW cascade. Figure 6.6 reproduces the longitudinal magnetoresistance, R_{xx}, and the transverse magnetoresistance, R_{yx}, of $(TMTSF)_2PF_6$ as reported by Kang [53].

Note the correspondence between the sharp increase in R_{xx} by a factor of more than one thousand and the decrease in the Hall resistance R_{yx} at about 22.5 T, indicating the emergence of the $N = 0$ ($n = 0$ in the P–H–T phase diagram) state, where in fact, a fractional quantum Hall effect (with $N = 1/m$) would be expected.

Another fascinating phenomenon was observed in the magnetoresistance measured as a function of the angle between the crystal axes and the magnetic field, as it rotates in a plane containing two crystal axes. Results obtained from $(TMTSF)_2PF_6$ under pressure [54, 55] showed that the magnetoresistance has dips at the so-called Lebed's magic angles [56, 57], given by

$$\tan\theta = \frac{p}{q}\frac{b\sin\gamma}{c\sin\beta\sin\alpha^*} - \cot\alpha^* \tag{6.2}$$

where p and q are small integers, and α, β, γ, b, c, and α^* are lattice parameters. Figure 6.7 shows results obtained for $(TMTSF)_2PF_6$ under a pressure of 7.9 kbar.

Several theories have addressed this phenomenon, but so far, none has been satisfactory. One of the possible explanations is based on the idea of coherence–incoherence transitions when a magnetic field is applied perpendicularly to the hopping direction [58].

6.1.5 From Fermi liquid to Luttinger liquid

As discussed near the end of 5.3 an efficient proof of Luttinger liquid behaviour in quasi-1D metals is the power law followed by transport properties such as the optical conductivity [16, 17] and the dc resistivity [59]. Figure 6.8 illustrates the scaling behaviour of the ac conductivity of some of the Bechgaard salts, including $(TMTSF)_2PF_6$.

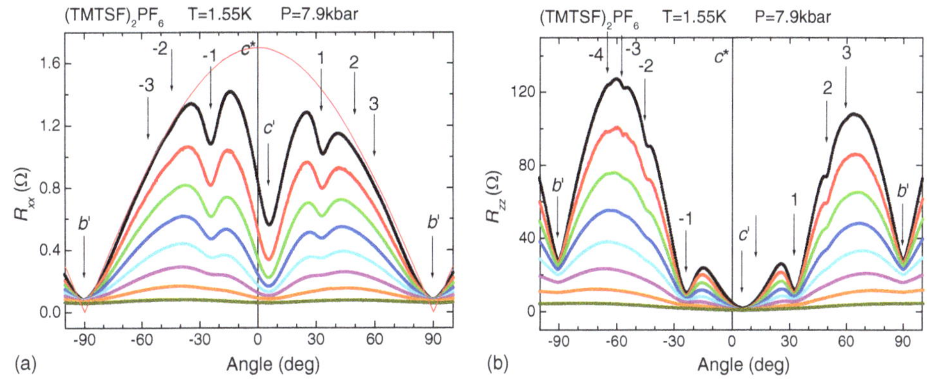

Figure 6.7. Angular magnetoresistances (a) R_{xx}, and (b) R_{zz} of $(TMTSF)_2PF_6$. The magnetic field is rotated in the $b'c^*$ plane. Arrows indicate the positions of the magic angles proposed by Lebed (the numbers correspond to the p/q ratio in equation 6.2). The fixed magnetic fields for each curve are 1–8 T in 1 T steps from lowest to highest. The thin red line in (a) represents an $R_{\|} \cos\theta$ fit of the 8 T data. Copyright (2004), reprinted from [53] with permission from Elsevier.

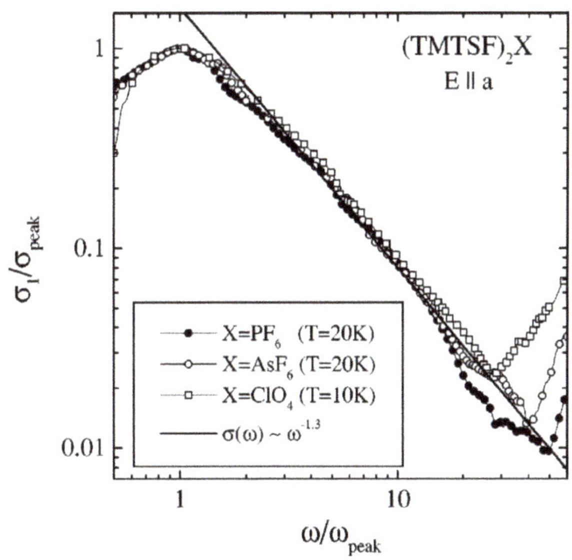

Figure 6.8. Normalized conductivities of $(TMTSF)_2PF_6$ and two other Bechgaard salts, shown on a log–log scale to demonstrate the power-law frequency dependence of the conductivities according to the Tomonaga–Luttinger liquid theory. The solid line is a fit for $\sigma(\omega) \sim \omega^{-\nu}$. For the three compounds, $\nu = 1.3 \pm 0.1$. Copyright 1998 by the American Physical Society, reprinted from [17, 18], with permission.

Given the experimental power law, $\sigma(\omega) \sim \omega^{-1.3}$, the common value of the Luttinger parameter $K_\rho \approx 0.23$ follows from equation (5.8), which is in excellent agreement with $K_\rho < 0.25$, the condition at which the quarter-filled Umklapp process becomes relevant. Similarly, consistent values of K_ρ and α were found from $\rho_{\|}(T)$ and $\rho_{\perp}(T)$ and $\sigma_{\|}(\omega)$ and $\sigma_{\perp}(\omega)$ [15, 59].

A wide range of a P–T phase diagram, such as the coloured area of figure 6.4, can be influenced by quantum critical points at which quantum phase transitions occur (at $T = 0$) as a result of competing ground states. In the case of $(TMTSF)_2PF_6$ the critical point is where the SDW is suppressed and superconductivity 'gradually' appears, and at higher pressure (above ≈20 kbar), the system becomes a two- or three-dimensional metal. This quantum critical point corresponds to the deconfinement transition (compare the phase diagrams of figures 5.7 and 6.4(b)). The deconfinement transition can take place when the interchain coupling reaches $t_\perp^*$ (figure 5.7), and charges can tunnel between stacks. As shown in the phase diagram (6.4(a)), when the temperature is increased, a dimensional crossover occurs from a two- or three-dimensional Fermi liquid to a Luttinger liquid. For $(TMTSF)_2PF_6$ the dimensional crossover is clearly seen at ambient pressure in the change of $\rho(T)$ at $T^* \approx 100$ K (see [22, 59, 60] for details).

6.1.6 Conclusions regarding $(TMTSF)_2PF_6$

We have tried to offer a general review of the most remarkable properties of the first reported organic superconductor. $(TMTSF)_2PF_6$ may not have been the much-sought high-temperature superconductor dreamt of in the 1970s, but it certainly became one of the most interesting solids due the richness of the physics it stimulated. Due to the quasi one-dimensional nature of its electronic structure, it became a prototype for the testing of sophisticated theories, such as the Tomonaga–Luttinger liquid theory, which predicts the properties of electrons interacting in one dimension. The facts that these organic conductors and superconductors are not strictly one dimensional, but that their one-dimensional chains interact through the interchain transfer integral, $t_\perp$, and that their Fermi surfaces are not really flat, but slightly warped in such a way that the effect of warping is sensitive to accessible temperatures, are gifts of nature to physicists, both theoreticians and experimentalists. These facts are at the roots of the mechanisms that control the instabilities and phase transitions observed in these solids.

It was, of course, impossible to even mention all the interesting properties and phenomena observed in $(TMTSF)_2PF_6$, but a search for titles containing $(TMTSF)_2PF_6$, for example, in the web of knowledge, will immediately reveal hundreds of articles describing new experiments and theoretical results.

6.2 The two-chain perylene–metal dithiolate series

Perylene–metal dithiolate salts have been synthesized in several stoichiometries, some of which have different crystal structures. An overview will be given of the α phases that have the general formula $(Per)_2[M(mnt)_2]$, where 'Per' denotes perylene, 'mnt' denotes maleonitriledithiolate, and M = Ni, Pd, Pt, Cu, Au, Co, Fe. We will focus on their structural, transport, and magnetic properties, low-temperature instabilities, and phase transitions, as well as their behaviour under extreme conditions of low temperatures, high pressures, and high magnetic fields. It is important to stress that the early experiments were performed using crystals of poor quality, or powders, due to difficulties in growing high-quality single crystals.

6.2.1 Structural features

The molecular structure of the perylene (Per) molecule, its HOMO, and the molecular structure of the metal-maleonitriledithiolate, $M(mnt)_2$ with the general formula $MS_4C_4(CN)_4$, where M=Ni, Pd, Pt, Cu, Au, Co, Fe, are shown in figure 6.9.

The $(Per)_2[M(mnt)_2]$ salts (where M=Ni, Pd, Pt, Cu, Au) form an isostructural family, with monoclinic crystal structures (space group $P2_1/n$) with $Z = 2$ at room temperature. In the Pt salt, $a = 16.612(4)$ Å, $b = 4.194(1)$ Å, $c = 30.211(6)$ Å, and $\beta = 118.70(2)°$ [61]. The other members of the family have very similar crystallographic parameters [62–64], with the exception of the Co and Fe salts, in which the b-axis is doubled and the metal–dithiolate units are dimerized [65]. The crystal structure of the Pt salt is represented in figure 6.10, which shows the projection along the conducting b-axis.

According to the stoichiometry and the crystallographic data, a more realistic formula for the family would be $(Per^{+0.5})_2[M(mnt)_2]^-$ (i.e. D_2A-type salts). The structural unit is asymmetrical and contains one $Per^{+0.5}$ cation and half a $[M(mnt)_2]^-$ anion, with the metal atom at the centre of symmetry (see figure 6.13(a) below). The unit cell contains four $Per^{+0.5}$ units and two $[M(mnt)_2]^-$ units related by a screw axis. The dihedral angle between the perylene and the $M(mnt)_2$ planes is 67°, and all

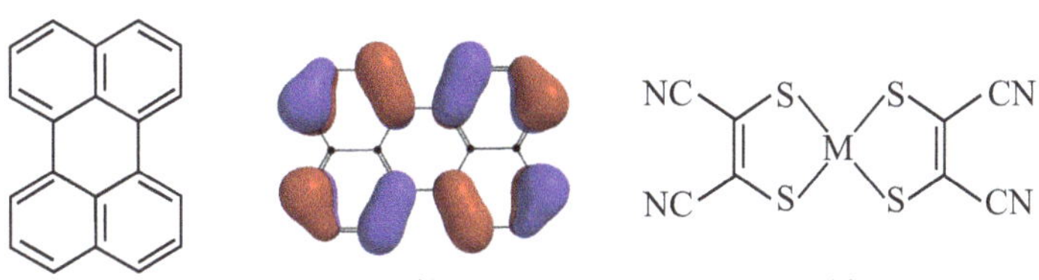

Figure 6.9. (a) Perylene molecule. (b) Perylene HOMO. (c) $M(mnt)_2$ molecule, M=Ni, Pd, Pt, Cu, Au, Co, Fe.

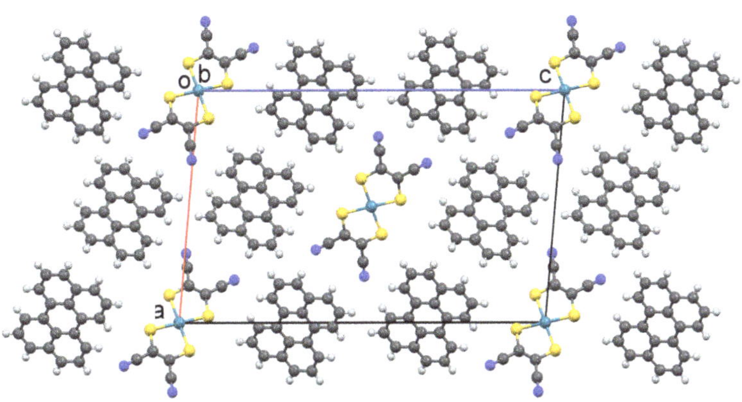

Figure 6.10. Room-temperature crystal structure of $(Per)_2[Pt(mnt)_2]$. Projection along the conducting b-axis showing segregated regular stacks of perylene and $Pt(mnt)_2$ parallel to b. Each perylene stack has three stacks of perylene and three stacks of $Pt(mnt)_2$ as first neighbours, and each stack of $Pt(mnt)_2$ is surrounded by six stacks of perylene. Reprinted from [61]. Copyright (1980) with permission from Elsevier.

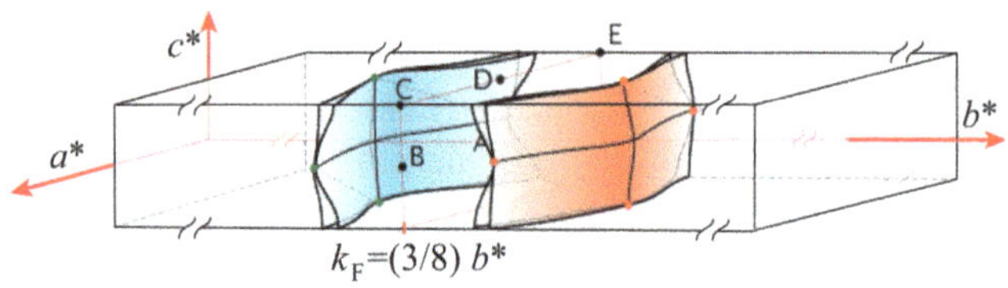

Figure 6.11. The four sheets of the $+k_F$ open Fermi surface for the four $Per^{+0.5}$ chain system in $(Per)_2[Pt(mnt)_2]$, highly enlarged along b^*. Reprinted from [66] by permission from Springer, copyright Springer, 2005.

interchain couplings are very weak. Along the perylene chains, the molecules, tilted by 37°with respect to *b*, overlap in an offset mode at a distance of 3.32 Å, as in graphite, but at a shorter distance (the distance in graphite is 3.354 Å). In the dithiolate chains, the $[M(mnt)_2]^-$ anions are tilted by 29° relative to *b* and separated by 3.65 Å. The adjacent Pt atoms are 4.19 Å apart.

The members of this family are two-chain systems in which both chains are active. As D_2A-type salts, each perylene chain is a three-quarters electron- (one-quarter hole-) filled band system. Each dithiolate chain is a half-filled band system when M=Ni, Pd, Pt, i.e. when the anion is magnetic ($S = 1/2$), and it is a full-band system when M=Cu or Au, i.e. when the anion is diamagnetic.

The band structure of the $(Per)_2[M(mnt)_2]$ salts was studied by Canadell *et al* [66]. The calculations made for the Pt salt (no significant differences are expected for the other salts) show that these salts are almost perfect one-dimensional conductors, and that conduction mainly occurs along the stacking *b*-axis of the $Per^{+0.5}$ chain, in agreement with the high experimental anisotropy [67]. The calculated transfer integrals were $t_b \approx 150$ meV, $t_a \approx 2$ meV and $t_c \approx 0$ meV, which are in good agreement with the experimental bandwidth, $4t \approx 0.4$–0.6 eV, depending on M, found in the thermopower and magnetic susceptibility data [64, 68, 69]. The band structure calculation was focused on the $Per^{+0.5}$ chain subsystem, since there was evidence that the dithiolate subsystem was insulating, and it was based on the room-temperature crystal structure of $(Per)_2[Pt(mnt)_2]$ shown in figure 6.10. With four perylene molecules per unit cell, there are four nearly degenerate HOMO bands along k_{b^*} (by convention, $(b \cdot b^*) = 2\pi$). If the transverse interactions could be ignored, the Fermi surface would be open and formed by four superposed planes at $k_F = \pm(3/8)b^*$. However, due to the small but non-negligible transverse interactions, the Fermi surface is split into four pairs of slightly warped sheets, as shown in figure 6.11.

6.2.2 Transport and magnetic properties

All members of the two-chain perylene–metal dithiolate family, $(Per)_2[M(mnt)_2]$, exhibit metallic transport behaviour at room temperature. The conductivities are of the order of 700 S cm^{-1} along the perylene chain (*b*-axis), with a measured anisotropy of the order of $\sigma_b/\sigma_a \approx 10^3$, when M=Au, Pt, Ni, Cu [67] (figure 6.12).

The conductivities of the other members (M=Pd, Co, Fe) are a little lower. All follow a metallic regime ($\rho \sim T^{1.5}$) down to a metal–insulator transition.

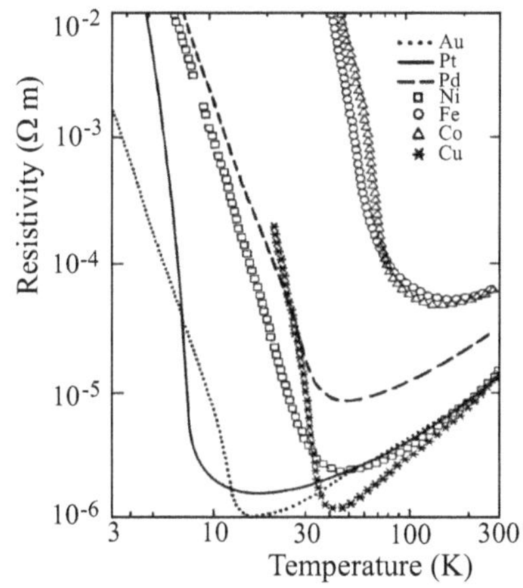

Figure 6.12. Temperature dependence of the electrical resistivities of the $(Per)_2[M(mnt)_2]$ salts (M=Au, Pt, Pd, Ni, Fe, Co, Cu). The conducting perylene chains undergo Peierls transitions to CDW ground states at low temperatures (Pt: 8.2, Au: 12, Ni: 25, Pd: 28, Cu: 33, Fe: 58, Co: 73 K) [70]. Reprinted by permission of Taylor and Francis Ltd.

The thermopower for a one-dimensional metal, in the tight-binding approximation (neglecting the energy-dependent scattering) is given by [71]:

$$S = -\frac{2\pi^2 k_B^2 T}{3eW}\left[\frac{\cos(\pi\rho/2)}{\sin^2(\pi\rho/2)}\right] \tag{6.3}$$

where $W = 4t$ is the bandwidth, e is the elementary electric charge, and ρ is the number of electrons per site. The room-temperature values of the thermopower lie in the range of 32 – 42 $\mu V K^{-1}$ for all members of the family, [68, 72, 73] and approximately follow equation (6.3) for $T > 100$ K. With $\rho = 3/2$, we can estimate a bandwidth (of the perylene chains) of $W \approx 0.6$ eV in agreement with the calculated transfer integrals t_b. At low temperatures the thermopower follows the $1/T$ behaviour typical of semiconductors. An important aspect of the thermopower results is that they not only give the bandwidths, but also the sign of the charge carriers which, as expected, is positive (holes).

Magnetic susceptibility measurements, electron-spin resonance (ESR) spectra, and their temperature dependence were of utmost importance in the earlier studies of this family of two-chain materials, indicating the presence of fast spin exchange between the localized spins of the magnetic $[Pd(mnt)_2]^-$ chains and the conduction electrons in the perylene chains [74–76]. In the presence of spin-exchange interactions between two species A and B, if the exchange rate is slow, the ESR spectrum will consist of two separate signals; if it is fast, there will be only one signal at an averaged position [77]. The two species will give a single narrow line if $J >> |g_A - g_B|\mu_B H$, where J is the exchange integral, μ_B is the Bohr magneton, and H is the magnetic field. In the present case, the averaged g value, $\bar{g}$, depends on the relative numbers of $[Pd(mnt)_2]^-$ and Per^+ species whose spins are interacting. If the fraction of interacting $[Pd(mnt)_2]^-$ spins is x, the g value of the averaged ESR line is $\bar{g} = g_{Per} + x(g_{Pd} - g_{Per})$. Using this rule, and combining the ESR and susceptibility results, the contributions to the susceptibility due to the localized electrons in the magnetic $[M(mnt)_2]^-$ (M=Ni, Pd, Pt) and the spins of the itinerant electrons in the perylene chains could be separated. The diamagnetic $[Au(mnt)_2]^-$ was used as a reference. At high temperatures, above the phase transitions, the susceptibilities of the Pd and Pt dithiolate chains follow Curie–Weiss behaviour, $\chi_{SP}^{-1} \propto (T - T_{SP})$. The results set the spin-Peierls transitions at $T_{SP} = 28$ K for the Pd salt, $T_{SP} = 7.5$ K for the Pt and $T_{SP} = 25$ K for the Ni analogues. These and other studies of the magnetic properties have been extended to all members of the family, and have been a centre of attention until the present [61, 67, 72, 74–76, 78].

6.2.3 Peierls and spin-Peierls instabilities and precursor effects

In the $(Per)_2[M(mnt)_2]$ series both the perylene and dithiolate chains are prone to the instabilities of one-dimensional interacting electron systems. Considering that the conducting perylene stacks form a three-quarters electron- (one-quarter hole-) filled band subsystem, one would expect CDW formation to take place at $2k_F \equiv 2k_F^{per} = b^*/4$ and a Peierls transition with tetramerization in the perylene chain. On the other hand, the dithiolate chains with M=Ni, Pd, Pt, in which the anions have spin 1/2, are half-filled band systems which can be Mott insulators and are expected to undergo spin-Peierls transitions with dimerization at $2k_F^{dit} = b^*/2$ (k_F^{dit} refers to the dithiolate subsystem). The M=Cu and Au compounds, whose anions are diamagnetic, correspond to full bands of the dithiolate chains, and are therefore expected to undergo CDW formation and Peierls transitions with tetramerization of the perylene chains. The Peierls and spin-Peierls transitions are indeed observed in the respective chains and, for each compound, both transition temperatures are the same, or very close, a feature which leads one to believe that they are coupled in some way. Figure 6.13 is an attempt to illustrate these facts.

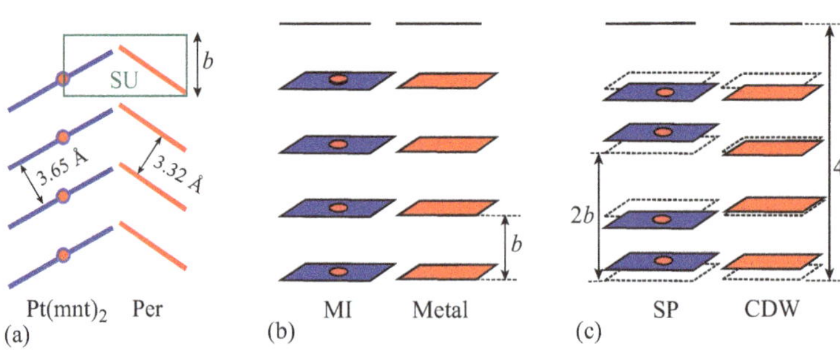

Figure 6.13. (a) $Pt(mnt)_2$ and Per chains along b. The top shows the structural unit (SU), which contains half a $[Pt(mnt)_2]^-$ and one $Per^{+0.5}$. (b) High-temperature arrangement of the chains (ignoring the tilt relative to b): the Mott-insulator (MI) $Pt(mnt)_2$ chain (left) and the metallic Per chain (right). (c) Ground state ($T = 0$) of the two chains: the dimerized spin-Peierls (SP) state of the $Pt(mnt)_2$ chain and the Peierls (CDW) state of the tetramerized perylene chain.

As mentioned in section 5.2, the one-dimensional instabilities lead to lattice fluctuations which can be observed by x-ray diffuse scattering measurements (see figure 4.4). Diffuse lines are expected to start appearing at temperatures $T_{\mathrm{P}}^{\mathrm{MF}}$, or $T_{\mathrm{SP}}^{\mathrm{MF}}$, predicted by mean-field theory, well above the metal–insulator transition temperatures, and condense into superlattice reflections at the Peierls or spin-Peierls transition temperatures, T_{P}, or T_{SP}, respectively.

In order to look for these precursor effects of the phase transitions, x-ray diffuse scattering measurements of the Pt, Pd, and Au salts were first made by Henriques *et al* [62]. One-dimensional structural fluctuations were observed in the metallic state of the Pd salt as diffuse lines at the $q_{\mathrm{SP}} = b^*/2$ wave vector, which corresponds to twice the Fermi wave vector of the dithiolate stacks, $2k_{\mathrm{F}}^{\mathrm{dit}}$, from 100 K down to the critical temperature, $T_{\mathrm{SP}}^{\mathrm{dit}} = 28$ K, at which superstructure reflections became well defined, as expected for a spin-Peierls transition. No diffuse lines were detected in the diamagnetic Au compound at temperatures down to the critical temperature $T_{\mathrm{c}} = 12$ K. However, a well-defined metal–insulator transition was later observed in this salt at 12.0 K (at zero magnetic field) with a field dependence consistent with the theoretical predictions for a Peierls transition [79]. Structural fluctuations were also observed in the Pt salt, in the form of diffuse lines at the wave vector $q = b^*/2$ which emerged at 25 K, well above the metal–insulator transition temperature $T_{\mathrm{c}} \approx 8$ K, which was not accessible in the experimental setup. However, the results of x-ray diffuse scattering measurements recently reported by Pouget *et al* [78] for the Pt compound show, at 10 K, intense diffuse lines at the $q_{\mathrm{SP}} = b^*/2$ wave vector, between the layers of the main Bragg reflections (see figure 4.4). A measurement at 4 K shows that these lines have condensed, not into well-defined superstructure reflections, but into broad diffuse spots, indicating that only a 3D SP short-range order is present below the SP transition temperature—the SP dimerization extends over 36 Å (i.e. 9b) in the stack direction and 16 Å in the transverse direction (i.e. the first-neighbour dithiolate distance). In addition, both $\chi_{\mathrm{SP}}^{-1}(q_{\mathrm{SP}})$ and the inverse coherence length, ξ_b^{-1}, tend to vanish at $\approx$7.5 K. These results establish that only a pseudo-SP transition takes place at $T_{\mathrm{SP}}' \approx 7.5$ K. Experimental data suggest that the $2k_{\mathrm{F}}^{\mathrm{dit}} = b^*/2$ pseudo-SP transition and the $2k_{\mathrm{F}}^{\mathrm{per}} = b^*/4$ Peierls transition in the Per chains might occur at slightly different temperatures [78–80], as mentioned above. Identical x-ray diffuse scattering measurements were performed by Gama *et al* [81], focusing on the metal–insulator transitions observed for Fe, Co, Ni, and Cu, at 58, 73, 25, and 33 K, respectively. The results show that the transitions in the Ni and Cu salts correspond to the tetramerization on the perylene sublattice and to its dimerization in the Fe and Co salts. Note that in both the Fe and Co salts, the b-axis is doubled at temperatures above the transition temperatures. Therefore, in all four cases, the distortions are associated with $2k_{\mathrm{F}} = b^*/4$ Peierls instabilities on the perylene chains. Furthermore, the magnetic chains of the Ni analogue undergo a $2k_{\mathrm{F}}^{\mathrm{dit}} = b^*/2$ distortion, with dimerization, ending up in a nonmagnetic spin-Peierls ground state.

Two important points about the instabilities in the Au compound should be stressed. First, the subtle feature of the Fermi surface (figure 6.11) with two pairs of warped sheets, combined with the opening of a Peierls gap ($2\Delta \approx 3.5$ meV) only slightly larger than the interchain transfer integral ($t_{\perp} \sim 2$ meV), is expected to weaken the Peierls instability under pressure, eventually leading to the suppression of the CDW and the generation of an SC state [78]. The second point is that the weak Peierls gap $2\Delta < \hbar\omega_c$ (ω_c is the critical phonon frequency) implies that the Peierls transition should occur in the non-adiabatic regime [82], in which quantum effects are relevant and expected to increase under pressure and lead to a vanishing CDW ground state that occurs at a quantum critical point (at $T = 0$).

6.2.3.1 Collective CDW transport

The Fröhlich theory of superconductivity in one-dimensional metals based on collective CDW transport [83] was not successful, but collective CDW (or SDW) transport does manifest itself in one-dimensional electron systems as nonlinear transport with some specific dynamics. In fact, when a dc voltage is applied to a one-dimensional system in the CDW state, the current may have two components: the current due to the electrons thermally excited across the Peierls gap, which vanishes at $T = 0$, and an oscillating component, with wavelength $\lambda_c = \pi/k_F = 4b$ (b is the lattice constant), due to sliding CDWs (in commensurate systems). This excess nonlinear conductivity is observed below the metal–insulator transition T_c and above a threshold field E_t determined by the impurity concentration. A typical feature of this process is the appearance, at the onset of nonlinearity, of narrowband noise with frequencies proportional to the excess current. In commensurate systems, the CDWs are pinned to the underlying lattice, and in incommensurate systems, they may be pinned to impurities and defects. Measurements performed at $T = 4.2$ K in the Au, Pt, and Co perylene–metal dithiolates [84–89] show all the main features of collective CDW nonlinear transport, in the CDW state, well below the metal–insulator transitions at T_c and above different threshold fields E_t. The dynamics of the CDW depinning process in the Au and Pt compounds are identical, the narrowband noise amplitude being weaker in the Au analogue, probably due to a larger amount of impurities or defects. For the Pt salt, Fourier analysis of the voltage noise at a constant dc current shows one fundamental frequency, which is proportional to the CDW current, and a large number of harmonics. A larger pinning energy (much higher E_t) observed for the Pt compound, compared with that of the Au compound, may reflect an additional commensurability effect due to the dimerization of the Pt dithiolate chains.

6.2.4 The nature of the instabilities in Ni, Pd, and Pt salts

The earlier work on $Per_2[Pd(mnt)_2]$ [74, 76] showing a single ESR line at a g value intermediate between those of $[Pd(mnt)_2]^-$ and Per^+ gives experimental evidence for a sizeable interstack exchange coupling $J_{\perp}$, or fast spin exchange between the conduction electrons and the localized spins on the Per and dithiolate stacks, respectively. Evidence of the same feature was provided by the low-temperature profile of the ESR linewidth of $Per_2[Pt(mnt)_2]$ combined with magnetic susceptibility data [90]. These results indicate that the two-chain $Per_2[M(mnt)_2]$ salts with M=Ni, Pd, and Pt are realizations of 1D Kondo lattices [91], where localized spins are

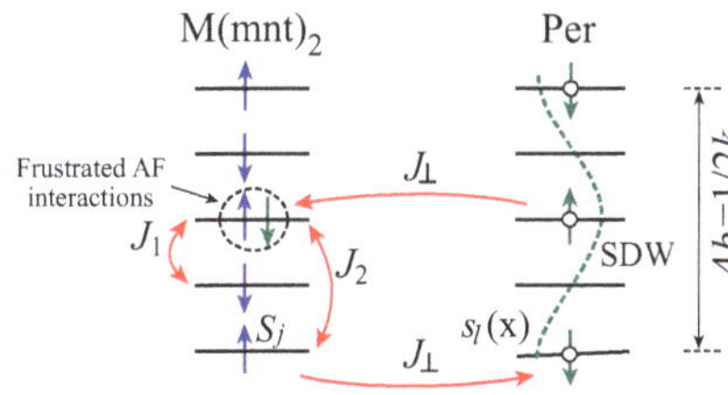

Figure 6.14. Representation of competing antiferromagnetic (AF) exchange interactions in the $Per_2[M(mnt)_2]$ salts for M = Ni, Pd, and Pt. $J_\perp$ is the transverse exchange coupling between the nearest dithiolate and Per spins S_j and $s_l(x)$. J_1 is the first-neighbour direct exchange interaction on the dithiolate stack, and J_2 is the second-neighbour indirect RKKY exchange interaction mediated by the induced $2k_F$ SDW on the Per stack. Reprinted with permission from [78], copyright 2017 by the authors.

coupled to the itinerant spins of a conducting electron gas. If many localized spins are arranged in a lattice, the same coupling constant J that induces the Kondo effect also induces a magnetic interaction between the localized spins. These localized spins can exchange, mediated by two itinerant spins (of the conducting electrons) that travel and scatter between the localized sites. This is therefore a long-range spin-exchange coupling involving two scattering events—the so-called RKKY interaction first considered by Ruderman and Kittel [92], Kasuya [93] and Yosida [94].

A careful examination of the crystal structure of the $Per_2[M(mnt)]_2$ salts (figure 6.10) shows that the antiferromagnetic interactions within the dithiolate stacks with M=Ni, Pd, and Pt could be mediated by the conducting Per stacks, in a similar way to the RKKY interaction, as suggested in [90, 95]. However, if the Per stacks are involved in such an RKKY type of interaction, one would expect to see induced $2k_F^{per} = b^*/4$ SDWs on the Per stacks. So far, there is no evidence of such SDWs, but the modulation of the Per stacks could be a mixed $2k_F^{per}$ SDW–CDW below T_P, as found in $(TMTSF)_2PF_6$ [8, 9]. This could happen, specifically in the case of the Pd salt, where no $2k_F^{per}$ Peierls superstructure reflections have been observed. The abovementioned spin-exchange interactions are schematically sketched in figure 6.14.

A spin S_j located on a dithiolate anion interacts with the spin density $s_l(x)$ located on a neighbouring Per stack, where it induces an SDW through the AF exchange coupling $J_\perp$. At a distance $2b$ (in this case), the oscillating spin density on the Per stack interacts, through $J_\perp$, with a new spin on the near neighbour dithiolate anion. Effective values of the intrastack AF interaction can be estimated from the susceptibilities, as follows: 100–300 K for M=Ni; 260 K above T = 100 K for M=Pd; and 35 K below T = 90 K for M=Pt.

There is a critical value of the frustration ratio, $\alpha_c = J_2/J_1 \approx 0.24$, below which the ground state of the system is a gapless spin fluid with a quasi long-range AF order. Although the SP transition for the Ni and Pd salts occurs in the adiabatic limit, the SP transition for the Pt salt occurs in the quantum limit and probably with $\alpha = J_2/J_1 < \alpha_c$ [78]. This makes the Pt salt particularly interesting and deserving of further investigation.

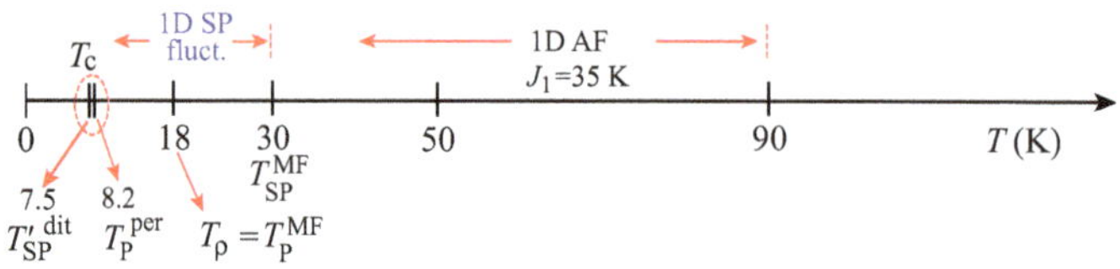

Figure 6.15. Relevant temperature scales for $(Per)_2[Pt(mnt)_2]$. One-dimensional AF interactions are observed below 90 K, and 1D SP fluctuations start at about 30 K. The $2k_F^{dit} = b^*/2$ pseudo-SP transition in the dithiolate stacks and the $2k_F^{per} = b^*/4$ Peierls transition in the Per stacks may occur at slightly different temperatures. RKKY interactions are expected below $T_\rho \approx 18$ K.

We now look in detail at the relevant temperatures of the $(Per)_2[Pt(mnt)_2]$ salt (figure 6.15).

The Per chain undergoes a Peierls transition at $T_P^{per} \approx 8.2$ K with tetramerization at $q_{per} = 2k_F^{per} = b^*/4$. Conversely, the Pt dithiolate chain undergoes a pseudo spin-Peierls transition at $T_{SP}^{'dit} \approx 7.5$ K with dimerization at $q_{dit} = 2k_F^{dit} = b^*/2$. The latter is viewed as a pseudo-SP transition since, due to the structural fluctuations starting at 30 K, the diffuse lines do not really condense into well-defined reflections below the transition temperature but only into broad diffuse spots, indicating that only a short-range order remains [78].

The instabilities of the two chains appear to be strongly coupled, since the Peierls and spin-Peierls transitions seem to coincide, occurring at the same or very close temperatures (see figure 6.15) as shown by the electrical transport, specific heat, and x-ray diffuse scattering results [78–80], in spite of the fact that their wave vectors are different although related ($q_{dit} = 2q_{per}$). Moreover, they remain coupled at high magnetic field strengths (up to 20 T), which is at variance with a simple mean-field description; symmetry breaking in the perylene CDW transition may drive the nearly simultaneous SP transition in the dithiolate chains [80]. A study of the alloys $(Per)_2[Pt_xAu_{(1-x)}(mnt)_2]$ also indicates that the metal–insulator transition is driven by the (dominant) Peierls instability, which in turn triggers the spin-Peierls transition [96].

The nature of the transitions has been thoroughly investigated, namely by ESR, resistivity, magnetic susceptibility, thermopower, x-ray diffuse scattering, as described above, and by specific heat, [97], and ^{1}H [90, 98] and ^{195}Pt [98] NMR relaxation ($1/T_1$). However, several questions remain, concerning the coupling mechanism of the SP and CDW transitions.

Fifty years have passed since the exchange interaction between the anion and the cation spins in the perylene–metal dithiolate salts was first noticed [74, 75]; the nature of the mechanism involved is still questioned and is now linked to the subsequently discovered Peierls and spin-Peierls instabilities. In fact, the RKKY coupling appears to be relevant to the magnetic $[M(mnt)_2]^-$ chains, as in the Pd salt, in which the RKKY interaction is expected to emerge at $T_\rho \approx 50$ K, and in the Pt salt, at $T_\rho \approx 18$ K [78]. Xavier *et al* [95] addressed the problem of dimerization induced by the RKKY interaction in a 1D Kondo lattice at one-quarter filling. Their results are relevant to $(Per)_2[M(mnt)_2]$ (M=Pt, Pd), but wanting—a two-chain theory is needed for systems in which the interacting spins S_j and the spin waves $s_l(x)$ are on different chains (figure 6.14).

6.2.5 $(Per)_2[M(mnt)_2]$ (M=Au, Pt) under strong magnetic fields

The first measurements of the effects of magnetic fields on the $(Per)_2[M(mnt)_2]$ salts were made by Bonfait *et al* [79] for the Au and Pt compounds. In their experiments, high-precision measurements of good-quality crystals clearly established the metal–insulator transition temperatures (under a zero magnetic field) at $T_c = 12.0 \pm 0.2$ K for the Au compound and $T_c = 8.20 \pm 0.02$ K for the Pt analogue. Measurements made under magnetic fields revealed, at 8 T, a shift of T_c from 12 K to 11.5 K, and from 8.2 K to 7.33 K, for the Au and Pt salts, respectively. The values of the shifts of T_c are in good agreement with the equation for the mean-field theory of the Peierls transition [99, 100]; the theory also predicts the temperature dependence of the magnetoresistance $\delta\rho$:

$$\frac{\delta T_c}{T_c(0)} = -\frac{\gamma}{4}\left(\frac{\mu_B B}{k_B T_c(0)}\right)^2, \qquad \frac{\delta\rho}{\rho} = -\frac{1}{2}\left(\frac{\mu_B B}{k_B T}\right)^2 + \ldots \tag{6.4}$$

where μ_B is the Bohr magneton, B is the magnetic field, and γ is a constant ≈ 1. For the spin-Peierls transition the mean-field theory equation is practically the same, but with $\gamma \approx 2$ [101]. Based on these arguments, it is reasonable to conclude that the metal–insulator transitions observed in the Au and Pt compounds are both Peierls-type transitions. Note that, as described above (see also figure 6.15) in the Pt salt, a spin-Peierls transition does occur in the dithiolate chain at nearly the same temperature as the Peierls transition in the perylene chain. Additional measurements of the Pt compound confirmed the coupling of the Peierls instability in the perylene chains to the spin-Peierls instability in the dithiolate chains [102], accompanied by complete suppression of both transitions at ca. 20 T as T approaches absolute zero [103]. In the Au compound, the Peierls transition is suppressed at 37 T as $T \to 0$ [100]. The results are in agreement with the theory, including the field dependence of the magnetoresistance and the energy gap. For fields tilted from the stacking axis, orbital effects appear above 15 T, which is consistent with an expected metallic phase beyond the CDW phase boundary. Figure 6.16 illustrates the general features of the phase diagrams of the Au and Pt salts and shows the SP, CDW, and *field-induced charge density wave* (FICDW) phases for magnetic fields of up to 45 T (measurements were made with fields parallel to a, b, and c and showed some anisotropy, which is ignored in this general picture).

As discussed above, the Pt salt exhibits a CDW ground state below 8.2 K coupled to a pseudo-SP at $\approx$7.5 K. Magnetoresistance and magnetization data show that the CDW is suppressed at magnetic fields of $\approx$20 T, above which, a cascade of FICDW subphases appears for all orientations of the magnetic field. Beyond this new state, a low-resistance state emerges at more than 40 T. These conclusions, supported by resistivity, thermopower, magnetization, and Hall-effect measurements, are discussed in detail by Graf *et al* [103] and Brooks *et al* [105, 106]. The two transitions (CDW and SP) in $(Per)_2[Pt(mnt)_2]$ appear to be coincident, but there is no clear

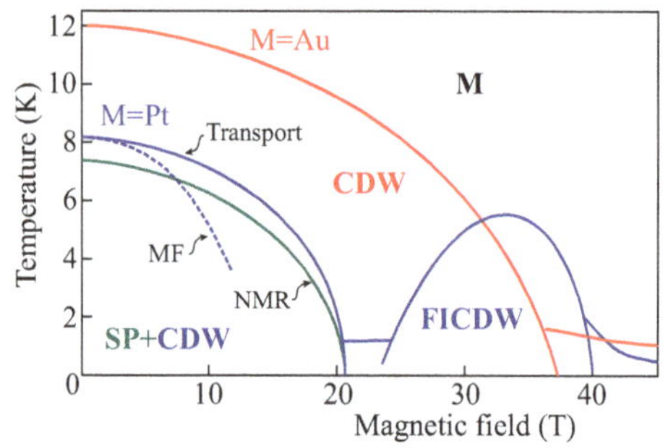

Figure 6.16. General features of the temperature vs magnetic field phase diagrams of $(Per)_2[M(mnt)_2]$ (M=Au, Pt) showing the boundaries of the CDW ($T_P^{per} = 12$ K) for M=Au (in red); and for M=Pt, the CDW ($T_P^{per} \approx 8.2$ K), the SP ($T_{SP}^{dit} \approx 7.5$ K), and the FICDW, derived from ^{1}H NMR and transport data, and the metallic (M) phases. 'MF' represents the mean-field predictions. (Based on data from [80, 98, 100, 102–105]).

theoretical explanation yet. Important progress has been made in understanding the magnetic FICDW phases, particularly by Lebed *et al* [107, 108]. A model for the electron spectrum of $(Per)_2[Pt(mnt)_2]$ has been suggested by Lebed *et al* [109], showing that spin-improved nesting effects may explain the emergence of the FICDW phase in high fields at relatively high temperatures (see figure 6.16), and predicting a phase transition to a periodic soliton wall superlattice (SWS) phase, a phase which has never been observed in solids.

6.2.6 $(Per)_2[M(mnt)_2]$ under pressure and strong magnetic fields

The effects of pressure on the transport properties of the Au, Pt [110], and Co [111] salts have shown that the evolution of electronic correlations with pressure is non-trivial, when compared with, for example, the $(TMTSF)_2X$ series. Additional studies have been performed to try to suppress the CDW using pressure and to study the effects of pressure and high magnetic fields on the CDW and FICDW states. Graf *et al* [112] reported that, at $T = 0.5$ K, the Au compound followed the mean-field predicted shift of T_{CDW} given by (6.4), where $\gamma \lesssim 1$, and the appearance of oscillations (which did not follow the typical Lifshitz–Kosevich formalism) reaching the $n = 1$ Landau level at $B \approx 14$ T for pressures of more than 4 kbar. Similar results have been reported for the Pt compound at $T = 0.7$ K and $B < 20$ T, but the FICDW persisted at higher fields for a pressure of 5.3 kbar.

Subsequent work with high-quality $(Per)_2[Au(mnt)_2]$ samples revealed that the oscillations observed in magnetoresistance at pressures at which the CDW transition is completely suppressed (5 kbar), for $B \| c$ and $I \| b$ (I is the current) were the result of the quantum interference of charge carriers in a magnetic field, which are distinct from the Shubnikov–de Haas oscillations associated with closed orbits and Landau quantization [113]. This effect, known as the Stark effect, is due to the interference of electron waves in metals discovered by Stark and Friedberg [114, 115]. It involves the interference of two electron trajectories on different Fermi surface sections, with a phase difference of $2\pi\phi/\phi_0$, where ϕ is the flux enclosed by the electron trajectories

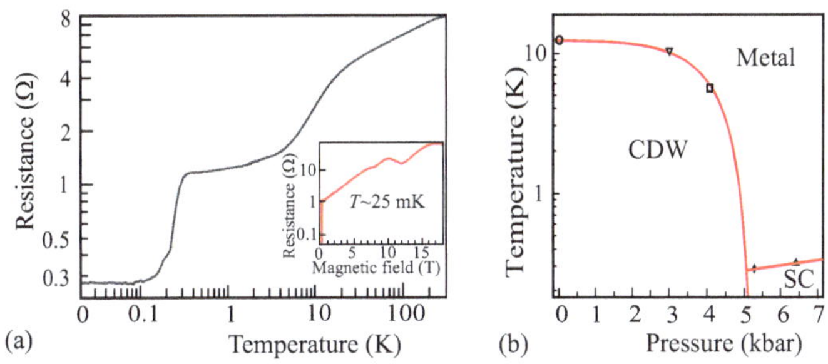

Figure 6.17. (a) Temperature dependence of the resistance in $(Per)_2[Au(mnt)_2]$ at 5.3 kbar indicating the onset of superconductivity at $T_c \approx 0.35$ K. The inset shows the magnetoresistance at 25 mK, the critical field (at $B_0 = 49.4$ mT), and the quantum interference effect. (b) Temperature–pressure phase diagram. Adapted from [117]. Reproduced with permission from EPL.

in k space, and ϕ_0 is the flux quantum h/e. The effect is observed in the form of large-amplitude oscillations, which are periodic with $1/B$, in the transverse magneto-resistance. The results are consistent with the calculated open-orbit Fermi surface topology [66] (figure 6.11). Apart from the quantum interference effects, geometrical effects associated with the Fermi surface topology have been independently identified in $(Per)_2[Au(mnt)_2]$ as angle-dependent magnetoresistance oscillations (AMRO) under constant magnetic fields (in the range of 6 to 31 T), at 5.4 and 6.4 kbar and $T = 300$ mK, for rotations in the *ac* and *bc* planes [116].

Superconductivity has been detected in the $(Per)_2[Au(mnt)_2]$ compound at a pressure of 5.3 kbar and $T_c = 0.313$ K by suppressing the CDW state [117]. Figure 6.17(a) represents the measured temperature dependence of the resistance for an optimal cooling rate of 0.5 K/min and a current of 5 μA. The resistance drops abruptly at $\approx$0.35 K, signalling the onset of superconductivity, but it does not completely vanish. The inset shows the magnetoresistance measured at 25 mK and indicates the critical field, established as being at $B_0 = 49.4$ mT, and the quantum interference effect oscillation.

Figure 6.17(b) outlines a general temperature–pressure phase diagram of this compound according to the available data. The CDW is completely suppressed at about 5 kbar. Above 3 kbar the magnetoresistance is typical of a metallic state in which orbital and quantum interference effects are observed. A systematic study of the resistance vs the field and temperature near the SC transition established the critical field (perpendicular to b) and temperature as $B_0 = 49.4$ mT and $T_c = 0.313$ K, respectively (see figure 3 of reference [117]). Measurements of the heat capacity and magnetic susceptibility were not attempted due to experimental difficulties.

It should be mentioned that in the Pt compound (with an $S = 1/2$ anion), an anomalously low resistance state has been observed near 20 T after the suppression of the main CDW (see figure 6.16) in several different samples and independent measurements [103, 106]. Although this may express the complete suppression of the CDW and a return to a new metallic state, the possibility of superconductivity should not be ruled out.

6.2.7 Conclusions regarding the perylene–metal dithiolate series

The $(Per)_2[M(mnt)_2]$ (M=Ni, Pd, Pt, Cu, Au, Co, and Fe) salts are highly one-dimensional. For the Pt compound, for example, the ratio of the transfer integral along the conducting axis and the larger perpendicular transfer integral is of the order of $t_{\|}/t_{\perp_1} \approx 75$, while for $TMTSF_2PF_6$ this ratio is $t_{\|}/t_{\perp_1} \approx 10$, and $t_{\|}/t_{\perp_2} \approx 33$. This important feature, and the fact that both the perylene and the dithiolate chains are active, one conducting, the other, either a Mott insulator (e.g. Ni, Pd, and Pt) or a common (full-band) insulator (e.g. Cu, Au), provide a huge diversity of physical states and properties, some of which we have addressed. There are certainly more unexpected results to be found as theoretical and experimental developments progress. And there are several unanswered questions, such as: what is the mechanism for the coupling of the SP and CDW instabilities? which instability drives the other? and what is the origin of the FICDW high-field phase in the Pt compound?

Further reading

Jerome D and Schulz H J 2002 *Adv. Phys.* 51 293–479 (https://doi.org/10.1080/00018730110116362)

Lebed A G (ed) 2008 *The Physics of Organic Superconductors and Conductors* (Berlin: Springer) (https://www.doi.org/10.1007/978-3-540-76672-8)

Mori T 2016 *Electronic Properties of Organic Conductors* (Japan: Springer)

Problems

6.1 The tight-binding equation (6.3) for the band structure contribution of the thermopower (S) in the metallic regime has been used for organic conductors. Consider the three-quarters filled band $(Per)_2[Au(mnt)_2]$ salt, with transfer integral $t = 0.145$ eV.

   (a) Calculate the thermopower of $(Per)_2[Au(mnt)_2]$ using equation (6.3), where the scattering contribution to S is negligible.

   (b) Are the charge carriers electrons or holes?

   (c) Now use equation (6.1) (as used for $(TMTSF)_2PF_6$) and compare the results.

   (d) For free electrons and energy-independent scattering, $S = -\pi^2 k_B^2 T/6eE_F$. Estimate the Fermi energy for $(Per)_2[Au(mnt)_2]$.

6.2 Why are some organic metals superconductors only under pressure?

References

[1] Bechgaard K, Jacobsen C S, Mortensen K, Pedersen H J and Thorup N 1980 *Solid State Commun.* **33** 1119–25

[2] Jérome D, Mazaud A, Ribault M and Bechgaard K 1980 *J. Phys. Lett.* **41** L95–98

[3] Medjanik K, Chernenkaya A, Kozina X, Nepijko S A, Öhrwall G, Foury-Leylekian P, Alemany P, Schönhense G, Canadell E and Pouget J P 2016 *J. Phys. Chem.* A **120** 8574–83

[4] Thorup N, Rindorf G, Soling H and Bechgaard K 1981 *Acta Cryst.* B **37** 1236–40

[5] Bourbonnais C and Jerome D 2008 Interacting electrons in quasi-one-dimensional organic superconductors *The Physics of Organic Superconductors and Conductors* ed A G Lebed (Berlin, Heidelberg: Springer) 357–412 ch 12

[6] Jacobsen C S, Tanner D B and Bechgaard K 1981 *Phys. Rev. Lett.* **46** 1142–5

[7] Andrieux A, Jérome D and Bechgaard K 1981 *J. Phys. Lett.* **42** 87–90

[8] Pouget J P and Ravy S 1996 *J. Phys. I France* **6** 1501–25

[9] Pouget J P and Ravy S 1997 *Synth. Met.* **85** 1523–8

[10] Andres K, Wudl F, McWhan D B, Thomas G A, Nalewajek D and Stevens A L 1980 *Phys. Rev. Lett.* **45** 1449–52

[11] Chaikin P M, Kwak J F, Jones T E, Garito A F and Heeger A J 1973 *Phys. Rev. Lett.* **31** 601–4

[12] Machida Y, Lin X, Kang W, Izawa K and Behnia K 2016 *Phys. Rev. Lett.* **116** 087003

[13] Ong N P, Wu W, Chaikin P M and Anderson P W 2004 *Europhys. Lett.* **66** 579–84

[14] Scott J C, Pedersen H J and Bechgaard K 1980 *Phys. Rev. Lett.* **45** 2125–8

[15] Moser J, Gabay M, Auban-Senzier P, Jérome D, Bechgaard K and Fabre J M 1998 *Eur. Phys. J.* B **1** 39–46

[16] Dressel M, Schwartz A, Grüner G and Degiorgi L 1996 *Phys. Rev. Lett.* **77** 398–401

[17] Schwartz A, Dressel M, Grüner G, Vescoli V, Degiorgi L and Giamarchi T 1998 *Phys. Rev.* B **58** 1261–71

[18] Vescoli V, Degiorgi L, Henderson W, Grüner G, Starkey K P and Montgomery L K 1998 *Science* **281** 1181

[19] Doiron-Leyraud N, Auban-Senzier P, Renéde Cotret S, Bechgaard K, Jérome D and Taillefer L 2010 *Physica B Condens. Matter* **405** S265–68

[20] Gor'kov L P and Grigoriev P D 2005 *Europhys. Lett.* **71** 425–30

[21] Greene R L and Engler E M 1980 *Phys. Rev. Lett.* **45** 1587–90

[22] Dressel M 2011 *J. Phys. Condens. Matter* **23** 293201

[23] Schulz H J, Jérome D, Ribault M, Mazaud A and Bechgaard K 1981 *J. Physique Lett.* **42** 51–4

[24] Azevedo L J, Schirber J E and Engler E M 1983 *Phys. Rev.* B **27** 5842–5

[25] Vuletić T, Auban-Senzier P, Pasquier C, Tomić S, Jérome D, Héritier M and Bechgaard K 2002 *Eur. Phys. J.* B **25** 319–31

[26] Tinkham M 1996 *Introduction to Superconductivity* (New York: McGraw-Hill)

[27] Lee I J, Chaikin P M and Naughton M J 2002 *Phys. Rev. Lett.* **88** 207002

[28] Lee I J, Brown S E, Yu W, Naughton M J and Chaikin P M 2005 *Phys. Rev. Lett.* **94** 197001

[29] Kang N, Salameh B, Auban-Senzier P, Jérome D, Pasquier C R and Brazovskii S 2010 *Phys. Rev.* B **81** 100509

[30] Pasquier C R, Kang N, Salameh B, Auban-Senzier P, Jérome D and Brazovskii S 2012 *Physica* **B407** 1806–9

[31] Brazovski S A, Gor'kov L and Lebed A G 1982 *Sov. phys. JETP* **56** 683-90 http://www.jetp.ras.ru/cgi-bin/dn/e_056_03_0683.pdf

[32] Gor'kov L P and Grigoriev P D 2007 *Phys. Rev.* B **75** 020507R

[33] Jérome D and Schulz H J 2002 *Adv. Phys.* **51** 293–479

[34] Jerome D 1985 *Mol. Cryst. Liq. Cryst.* **119** 1–9

[35] Lee I J, Brown S E, Clark W G, Strouse M J, Naughton M J, Kang W and Chaikin P M 2001 *Phys. Rev. Lett.* **88** 017004

[36] Fulde P and Ferrell R A 1964 *Phys. Rev.* **135** A550–63
[37] Fuseya Y, Bourbonnais C and Miyake K 2012 *Europhys. Lett.* **100** 57008
[38] Miyawaki N and Shimahara H 2014 *J. Phys. Soc. Japan* **83** 024703
[39] Podolsky D, Altman E, Rostunov T and Demler E 2004 *Phys. Rev. Lett.* **93** 246402
[40] Nickel J C, Duprat R, Bourbonnais C and Dupuis N 2005 *Phys. Rev. Lett.* **95** 247001
[41] Cooper J R, Kang W, Auban P, Montambaux G, Jérome D and Bechgaard K 1989 *Phys. Rev. Lett.* **63** 1984–7
[42] Hannahs S T, Brooks J S, Kang W, Chiang L Y and Chaikin P M 1989 *Phys. Rev. Lett.* **63** 1988–91
[43] Kang W, Hannahs S T and Chaikin P M 1993 *Phys. Rev. Lett.* **70** 3091–4
[44] Kornilov A V, Pudalov V M, Kitaoka Y, Ishida K, Mito T, Brooks J S, Qualls J S, Perenboom J A A J, Tateiwa N and Kobayashi T C 2002 *Phys. Rev.* B **65** 060404
[45] Kornilov A V, Pudalov V M, Kitaoka Y, Ishida K, Mito T, Brooks J S and Qualls J S 2003 *Synth. Met.* **133-34** 69–70
[46] Kornilov A V, Pudalov V M, Kitaoka Y, Ishida K, Zheng G Q, Mito T and Qualls J S 2004 *Phys. Rev.* B **69** 224404
[47] Poilblanc D, Montambaux G, Héritier M and Lederer P 1987 *Phys. Rev. Lett.* **58** 270–3
[48] Héritier M 2008 Cascade of FISDW Phases: Wave Vector Quantization and its Consequences *The Physics of Organic Superconductors and Conductors* (Springer Series in Materials Science vol 110) ed A G Lebed (Berlin, Heidelberg: Springer) pp 41–5
[49] Gor'kov L and Lebed A 1984 *J. Physique Lett.* **45** 433–40
[50] Héritier M, Montambaux G and Lederer P 1984 *J. Phys. Lett.* **45** 943–52
[51] Yamaji K 1985 *J. Phys. Soc. Japan* **54** 1034–40
[52] Azbel M Y, Bak P and Chaikin P M 1986 *Phys. Lett.* A **117** 92–6
[53] Kang W 2004 *Curr. Appl. Phys.* **4** 263–6
[54] Kang W, Hannahs S T and Chaikin P M 1992 *Phys. Rev. Lett.* **69** 2827–30
[55] Lee I J and Naughton M J 1998 *Phys. Rev.* B **58** R13343 46
[56] Lebed A G 1986 *JETP Lett.* **43** 137 http://jetpletters.ru/ps/1401/article_21263.pdf
[57] Lebed A G and Bak P 1989 *Phys. Rev. Lett.* **63** 1315–7
[58] Strong S P, Clarke D G and Anderson P W 1994 *Phys. Rev. Lett.* **73** 1007–10
[59] Dressel M, Petukhov K, Salameh B, Zornoza P and Giamarchi T 2005 *Phys. Rev.* B **71** 075104
[60] Pashkin A, Dressel M, Hanfland M and Kuntscher C A 2010 *Phys. Rev.* B **81** 125109
[61] Alcácer L, Novais H, Flandrois S, Coulon C, Chasseau D and Gautier J 1980 *Solid State Commun.* **35** 945
[62] Henriques R T, Alcácer L, Pouget J P and Jerome D 1984 *J. Phys. C Solid State Phys.* **17** 5197–208
[63] Domingos A, Henriques R T, Gama V, Almeida M, Vieira A L and Alcácer L 1988 *Synth. Met.* 411–6
[64] Gama V, Almeida M, Henriques R T, Santos I C, Domingos A, Ravy S and Pouget J P 1991 *J. Phys. Chem.* **95** 4263–7
[65] Gama V, Henriques R T, Almeida M and Alcácer L 1991 *Synth. Met.* **42** 2553–6
[66] Canadell E, Almeida M and Brooks J S 2004 *Eur. Phys. J.* B **42** R453
[67] Alcácer L 1985 *Mol. Cryst. Liq. Cryst.* **120** 221–8
[68] Henriques R T, Almeida M, Matos M J, Alcácer L and Bourbonnais C 1987 *Synth. Met.* **19** 379–84

[69] Almeida M and Henriques R 1997 Conducting Polymers *Organic Conductive Molecules and Polymers* vol 1 (Chichester: Wiley) 2 87–149
[70] Gama V, Henriques R T, Bonfait G, Almeida M, Ravy S, Pouget J P and Alcácer L 1993 *Mol. Cryst. Liq. Cryst.* **234** 171–8
[71] Chaikin P M, Greene R L, Etemad S and Engler E 1976 *Phys. Rev.* B **13** 1627–32
[72] Henriques R T, Alcácer L, Almeida M and Tomic S 1985 *Mol. Cryst. Liq. Cryst.* **120** 237–41
[73] Gama V, Henriques R T and Almeida M 1989 *Fizika* **21** 19–22
[74] Alcácer L 1970 Highly conductive molecular complexes *PhD thesis* (Riverside: University of California)
[75] Alcácer L and Maki A H 1974 *J. Phys. Chem.* **78** 215–7
[76] Alcácer L and Maki A H 1976 *J. Phys. Chem.* **80** 1912–6
[77] Pople J A, Schneider W G and Bernstein H J 1959 *High Resolution Nuclear Magnetic Resonance* (New York: McGraw-Hill) 218 p
[78] Pouget J P, Foury-Leylekian P and Almeida M 2017 *Magnetochemistry* **3** 13
[79] Bonfait G, Lopes E B, Matos M J, Henriques R T and Almeida M 1991 *Solid State Commun.* **80** 391–4
[80] Green E L, Lumata L L, Brooks S J S, Kuhns P L, Reyes A P, Almeida M, Matos M J, Henriques R T, Wright J A and Brown S E 2011 *Phys. Rev.* B **84** 121101 (R)
[81] Gama V, Henriques R T, Almeida M and Pouget J P 1993 *Synth. Met.* **55-57** 1677–82
[82] Caron L G and Bourbonnais C 1984 *Phys. Rev.* B **29** 4230–41
[83] Frohlich H 1954 *Proc. R. Soc. A* **223** 296
[84] Lopes E B, Matos M J, Henriques R T, Almeida M and Dumas J 1994 *Europhys. Lett.* **27** 241–6
[85] Lopes E B, Matos M, Henriques R T, Almeida M and Dumas J 1995 *Synth. Met.* **70** 1267–70
[86] Lopes E B, Matos M J, Henriques R T, Almeida M and Dumas J 1995 *Phys. Rev.* B **52**
[87] Dumas J, Thirion N, Lopes E B, Matos M J, Henriques R T and Almeida M 1995 *J. Phys. I France* **5** 539–45
[88] Lopes E, Matos M, Henriques R, Almeidal M and Dumas J 1996 *J. Phys.* I **6** 2141–9
[89] Lopes E B, Matos M J, Henriques R T, Almeida M and Dumas J 1997 *Synth. Met.* **86** 2163–4
[90] Bourbonnais C, Henriques R T, Wzieteck P, Kongle D, Voiron J and Jérome D 1991 *Phys. Rev.* B **44** 641–51
[91] Kondo J 1964 *Prog. Theor. Phys* **32** 37–49
[92] Ruderman M A and Kittel C 1954 *Phys. Rev.* **96** 99–102
[93] Kasuya T 1956 *Prog. Theor. Phys.* **16** 45–57
[94] Yosida K 1957 *Phys. Rev.* **106** 893–8
[95] Xavier J C, Pereira R G, Miranda E and Affleck I 2003 *Phys. Rev. Lett.* **90** 247204
[96] Matos M, Gama V, Bonfait G and Henriques R 1993 *Synth. Met.* **56** 1858–63
[97] Bonfait G, Matos M J, Henriques R T and Almeida M 1993 *J. Phys. IV France* 251–4
[98] Green E L, Lumata L, Brooks J S, Kuhns P, Reyes A, Brown S E and Almeida M 2012 *Crystals* **2** 1116–35
[99] Tiedje T, Carolan J F, Berlinsky A J and Weiler L 1975 *Can. J. Phys.* **53** 1593–605
[100] Graf D, Brooks J S, Choi E S, Uji S, Dias J C, Matos M and Almeida M 2004 *Phys. Rev.* B **69** 125113

[101] Bray J W, Interrante L V, Jacobs I S and Bonner J C 1983 The spin-Peierls transition *Extended Linear Chain Compounds* ed T S Miller (New York and London: Plenum Press) 377 p

[102] Matos M, Bonfait G, Henriques R T and Almeida M 1996 *Phys. Rev.* B **54** 15307–13

[103] Graf D, Choi E S, Brooks J S, Matos M, Henriques R T and Almeida M 2004 *Phys. Rev. Lett.* **93** 076406

[104] Graf D, Choi E S, Brooks J S, Dias J C, Henriques R T, Almeida M, Matos M and Rickel D 2005 *Synth. Met.* **153** 361–4

[105] Brooks J S, Graf D, Choi E S, Almeida M, Dias J C, Henriques R T and Matos M 2006 *J. Low Temp. Phys.* **142** 787–803

[106] Brooks J, Graf D, Choi E, Almeida M, Dias J, Henriques R and Matos M 2006 *Curr. Appl. Phys.* **6** 913–8

[107] Lebed A G 2003 *JETP Lett.* **78** 138–42

[108] Lebed A G 2009 *Phys. Rev. Lett.* **103** 046401

[109] Lebed A G and Wu S 2007 *JETP Lett.* **86** 135–8

[110] Henriques R T, Alcácer L, Jerome D, Bourbonnais C and Weyl C 1986 *J. Phys. C Solid State Phys.* **19** 4663

[111] Almeida M, Gama V, Santos I C, Graf D and Brooks J S 2009 *Cryst. Eng. Comm.* **11** 1103–8

[112] Graf D, Choi E S, Brooks J S and Almeida M 2006 *J. Low Temp. Phys.* **142** 179–84

[113] Graf D, Brooks J S, Choi E S, Almeida M, Henriques R, Dias J C and Uji S 2007 *Phys. Rev.* B **75** 245101 08

[114] Stark R W and Friedberg C B 1971 *Phys. Rev. Lett.* **26** 556–9

[115] Sandesara N B and Stark R W 1984 *Phys. Rev. Lett.* **53** 1681–3

[116] Graf D, Brooks J S, Choi E S, Almeida M, Henriques R T, Dias J C and Uji S 2009 *Phys. Rev.* B **80** 155104 08

[117] Graf D, Brooks J S, Almeida M, Dias J C, Uji S, Terashima T and Kimata M 2009 *Europhys. Lett.* **85** 27009

Chapter 7

Organic semiconductors for organic electronics

Representative organic semiconductors used in organic electronics are described in this chapter. They include π-conjugated organic molecules and polymers. The emphasis is on their structures, orbitals, bands, frontier orbitals (highest occupied molecular orbital (HOMO) and lowest unoccupied molecular orbital (LUMO)), topological defects (solitons and polarons), excitons, and transport and optical properties. Relevant aspects of the spectroscopy and photophysics of organic semiconductors are also introduced.

7.1 Introduction

The real interest in organic semiconductors as materials for electronics may be traced back to the mid-1980s, when the first reports were published of organic electronic devices with potential technological impact. The first organic solar cell with a significantly high fill factor, of a new type based on a molecular junction, was disclosed by Ching Tang at Kodak in 1986 [1]. This was closely followed by reports of the first polymer field-effect transistor by Tsumura, Koezuka, and Ando at Mitsubishi in Japan [2], and the first organic light-emitting diode (OLED) by Tang and van Slyke in 1987 [3]. In 1990, Richard Friend and co-workers at Cambridge reported the discovery of electroluminescence in conjugated polymers and the first polymer OLED (or PLED) [4]. In the following years, various conjugated polymers were applied as materials for polymer LEDs [5–7]. Other important milestones were the independent introductions by Yu *et al* [8] and Halls *et al* [9] of the concept of the bulk heterojunction (BHJ), a phase-separated polymer blend for solar cells, in 1995 and the discovery of highly efficient phosphorescent emission from electroluminescent devices in 1998 [10].

It should be mentioned that in 1950, Hiroo Inokuchi and co-workers discovered semiconducting properties in conjugated organic molecules and correctly explained their electrical properties in terms of π electrons [11] and also that the first two-layer organic solar cell was reported by Kearns and Calvin in 1958 [12]. And, of course,

doi:10.1088/978-0-7503-3347-4ch7

there was the discovery in 1977 of the high conductivity of doped polyacetylene [13], which triggered great interest in the study of conjugated polymers as materials of technological potential. Typical representatives of such polymers are polypyrrole [14, 15], poly(p-phenylene vinylene) [16], polythiophene [17], polyaniline [18], and soluble polythiophene and polypyrrole derivatives [19].

In the last twenty years there has been an increasing interest in the applications of organic semiconductors, such as OLED displays for smart phones, television screens, and solid-state lighting; and lately organic solar cells (OSCs), which are among the most promising candidates for next-generation solar cells. Electronic circuits, sensors, and drug delivery systems based on organic semiconductors will most probably bring great benefits to society in the near future, as a new organic bioelectronics technology emerges. All these technologies benefit from the immense variety of materials that synthetic chemistry can create. Organic electronics is an interdisciplinary field, which requires experts in many fields of science and technology who have a working understanding of the whole field. Those working in this field should have an open mind and be prepared for the discovery of applications which are not those that one could expect, as has happened in this particular field many times in the past.

The organic semiconductors used in organic electronics are either small molecules or conjugated polymers. Both of these generally form amorphous or polycrystalline films, but while small molecules are deposited from the vapour phase or from solution, polymers are always deposited from solution or formed *in situ* by electrochemical synthesis. Organic semiconductors mainly differ from conventional inorganic semiconductors (e.g. Si, Ge, and GaAs) by their type of structure, transport properties, and photophysical and optical properties. While in inorganic semiconductors the atoms are linked by covalent (or mixed covalent and ionic) bonds, the molecules or polymer chains in organic semiconductors are held together by van der Waals interactions, and eventually by hydrogen bonds; the molecules or polymer chains are highly conjugated. This makes organics 'soft materials', with a relatively high degree of dynamic and static disorder, leading to optical and electronic properties that are unique and distinct from those of inorganic semiconductors. A thin film or a single crystal of an organic semiconductor is, in fact, an arrangement of interacting molecules.

In particular, due to the type of chemical bond, the bandwidths of organics are of the order of 0.01–0.5 eV, compared to 1–4 eV in inorganics, and consequently the charge mobilities are usually $\leqslant 1\ \mathrm{cm^2\ V^{-1}\ s^{-1}}$ in organics vs $\sim 1000\ \mathrm{cm^2\ V^{-1}\ s^{-1}}$ in inorganics. This is one of the major differences that determines and limits the possible applications of organic electronics. While band theory may be very useful, many properties of organic semiconductors are better described in terms of the *frontier orbitals*, i.e. the HOMO and the LUMO. The HOMO level corresponds to the top of the valence band (VB) and the LUMO to the bottom of the conduction band (CB). In organic semiconductors, charge transport is described in terms of polaron hopping, a polaron being a quasiparticle formed by a charge and the associated lattice deformation. Of particular importance for OLEDs and OSCs are excitons of the Frenkel or charge-transfer type, with binding energies in the range of ~0.5–1 eV. And, of course, organic devices can be flexible, and in some cases, transparent.

We tend to associate organic electronics with semiconductors alone, as in conventional electronics, but as we discuss in other chapters of this book, there are organic metals and superconductors which are charge-transfer salts, and highly conductive polymers such as polyaniline and polypyrrole. These polymers are widely used in many applications. For organic electronics, one of the most popular conductive polymers is PEDOT:PSS (see the section Identification of chemical names mentioned in the text for chemical name definitions), because it is transparent, highly conductive, and commercially available in the form of a colloidal aqueous dispersion that is easy to process.

7.2 Molecular, crystal, and electronic structures

The main feature that explains significant electron conduction in organic molecular materials is the existence of bond conjugation in the chemical structure. This means that there are sequences of alternate double and single carbon–carbon bonds which allow electron delocalization. This structural feature, which can be named a 'π-system', is a consequence of the electronic configuration of the carbon atom, $1s^2 2s^2 2p^2$, where the four valence orbitals ($2s$ and $2p$) form an sp^2 hybrid orbital set that provides three σ bonds, and the p_z orbital perpendicular to the sigma bond plane forms a π-bond.

To illustrate this feature we choose the simple model of a polyacetylene[1] chain in which a regular linear lattice contains one -CH=CH- unit per site. In the polyacetylene chain, the three σ bonds form the backbone of the chain; two connect the adjacent carbon atoms, and the third forms the bond with the hydrogen atom. The p_z orbital of each carbon combines with its neighbours to form a bonding π orbital (HOMO) and an anti-bonding π^* orbital (LUMO). In the context of band theory, which is appropriate for a very large number of -CH=CH- units, the π and π^* energy levels broaden and form the VB and CB, respectively. The HOMO and the LUMO roughly correspond to the top of the VB and the bottom of the CB, respectively (figure 7.1).

It may, at first sight, seem reasonable to consider a polyacetylene chain with equal bond lengths in a linear lattice with one CH unit per site. In a band structure calculation for that model, the σ levels would give a full σ band and the p_z orbital would give a half-filled π band, since there would be only one p_z orbital, with one electron, per CH unit. One would then expect polyacetylene to be a metal. In fact, various studies of the crystal structure of polyacetylene have indicated that there are not one, but two -CH=CH- units per primitive cell [20]. This implies (for the valence orbitals only) the existence of two pairs of nearly degenerate bands separated by a gap. The existence of alternate bond lengths in polyacetylene is, however, a matter of controversy [21]. It should also be noted that high degrees of doping affect both the crystal and electronic structures, and there are a large number of defects.

[1] Unless specified otherwise, when mentioning polyacetylene, we refer to trans-polyacetylene.

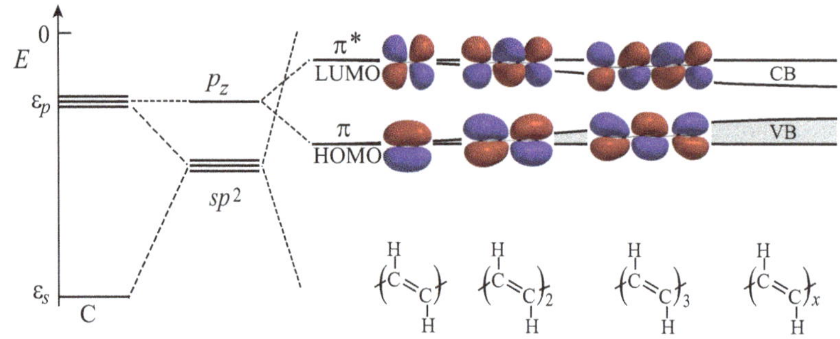

Figure 7.1. As a result of the formation of sp^2 and p_z orbitals in a C atom, the combination of neighbouring p_z orbitals allows the formation of a bonding π orbital (HOMO) and an anti-bonding π^* orbital (LUMO) in a -CH=CH- unit. A polyacetylene chain can, in principle, be formed by adding -CH=CH- units, and the corresponding π and π^* energy levels lead to a VB and a CB, respectively, for a large number, x, of units. The bands are formed by linear combinations of the primitive cell orbitals.

For a small-molecule-based semiconductor, such as pentacene, the same reasoning applies. The HOMO and LUMO orbitals are built in the same way in the π-systems, which are always present in the molecular structure. Detailed knowledge of the frontier orbitals and levels is of the utmost importance when determining the relevant properties for any application. The molecular structures of both the small molecules and the repeating polymer units are therefore major concerns when designing a new material, and it pays to combine chemical intuition and knowledge with predictive computational design based on quantum chemical calculations, molecular dynamics methods, or any other simulation approach to predict not only the equilibrium molecular geometry and electronic structure, but also the possible structural arrangement. An important aspect in which theory can be very useful to guide the organic synthesis is the prediction of how different chemical functional groups modulate the electronic, optical, and other properties, such as solubility.

Chemists frequently use computational methods based on the Hartree–Fock (HF) approximation and on density functional theory (DFT) to predict energy levels and orbitals in molecules, in particular the HOMO and LUMO, and their closest neighbours among the occupied and unoccupied orbitals. One of the most popular methods is the B3LYP/6-31G* version of DFT, where B3LYP stands for Becke, 3-parameter, Lee–Yang–Parr, and 6-31G* is the basis set.

From these calculations one can also obtain the most probable geometry of the molecule, along with several parameters and property maps, such as the electron density maps. The ionization potential, for example, is the negative of the energy of the HOMO level ($IP = -\varepsilon_{\text{HOMO}}$) in the HF approximation, according to Koopmans' theorem, and the first electron affinity is given by $EA = -\varepsilon_{\text{LUMO}}$. In DFT the ionization potential is also given by $IP = -\varepsilon_{\text{HOMO}}$. These methods are explained in some detail in our book suggested in the Further reading section at the end of this chapter.

The most common parameters that characterize organic semiconductors are schematically represented in figure 7.2.

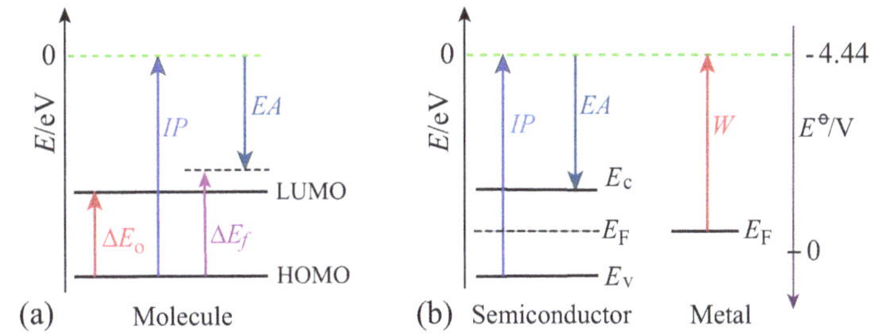

Figure 7.2. (a) The ionization potential (*IP*), electron affinity (*EA*), optical gap (ΔE_o), and fundamental gap (ΔE_f) in molecules. (b) *IP* and *EA* in semiconductors and the work function (*W*) in metals. E_c denotes the bottom of the conduction band, E_F denotes the Fermi energy, and E_V denotes the top of the valence band. The right-hand side shows the electrochemical scale of reduction potentials (in volts) relative to the standard hydrogen electrode, for comparison.

In a molecule, the energy of the lowest electronic transition is the *optical gap*, ΔE_o. The first ionization potential is the energy required to remove an electron from the HOMO level of an isolated molecule; the first electron affinity, *EA*, is the negative of the energy change caused by adding an electron to the molecule; and the difference, $\Delta E_f = IP - EA$, is the *fundamental gap*, which is only approximately given by the difference between the LUMO and HOMO levels, this being the *HOMO–LUMO gap*. The optical gap is smaller than the fundamental gap due to the fact that in the excited state, the electron–hole pair generated by the optical transition has a binding energy $E_b = \Delta E_f - \Delta E_o$. Note that the HOMO and LUMO are one-electron levels, not many-electron states, and therefore the calculated values (which may depend on the method used) differ from the experimental energies needed to remove one electron from a many-electron molecule (the *IP*) or to add one electron to a molecule (the *EA*). For those reasons, values obtained from the literature should be taken as indicative only.

The corresponding values for a solid are shown in the energy band schematic in (b) for a semiconductor and for a metal. The electron affinity is the change in energy when one electron is moved from the vacuum level to the bottom of the conduction band. The bandgap is the difference between the bottom of the conduction band, E_c, and the top of the valence band, E_v. We can also define a *transport gap*, E_g, as the minimum energy needed to create a charge carrier. Another important parameter is the *work function*, *W*, which is of particular importance for metals, being the energy barrier that keeps the electrons in the solid. At the far right is the electrochemical scale of standard reduction potentials, E^0, in volts, for comparison with the energy scale, which is shown in electronvolts.

The meanings of the various parameters, particularly the energy gaps, which can be measured or calculated are discussed in a short article by Jean-Luc Brédas [22], and a review of the experimental methods and corrections that should be used is given by Sworakowski [23]. A popular method used to determine the HOMO and LUMO energies is cyclic voltammetry, but the values obtained by this method for small molecules are generally lower by 16% than the 'true' values (the energies of the transport gaps as measured by UV photoelectron spectroscopy (UPS) and inverse photoelectron spectroscopy (IPES)).

In organic electronics most materials are used in the form of thin films deposited on substrates. The structure and morphology of the films depend on the type of substrate, on the type of interaction between the materials and the substrates, and on the deposition technique, i.e. whether it is based on the vapour phase or a solution or the use of laboratory or industrial printing techniques. Using small molecules and appropriate substrates, one may achieve a kind of epitaxial or quasiepitaxial growth, taking advantage of the fact that the bonding interactions between the molecules being deposited are of the weak van der Waals type, which allows some degree of freedom for the growth of the first layers to adjust to the structure of the substrate; the subsequent growth of additional layers forms ordered molecular thin films or equilibrial crystal structures, which are eventually different from bulk single crystals. Polymer thin films are generally deposited from a solution or a suspension, and their quality can be improved by post-deposition treatments, such as thermal and solvent annealing. The growth rate of both polymer and small-molecule films is critical for device performance. Self-assembly and self-organization are also very important for the formation of high-quality films. They are particularly relevant for phase separation in the polymer blends used to make bulk heterojunction solar cells.

There are thousands of organic semiconductors, especially photoactive polymers for organic solar cells. This is an advantage, since we have many choices of material for each specific device type, but the time will come when technologies will concentrate on a few building-block materials in order to allow for mass production and lower costs. The great success of today's low-cost electronics is, in part, the result of concentrating on a few materials, such as silicon, which has been perfected to an extreme extent. The great advantage of variety and diversity is the potential emergence of new and even unexpected applications, particularly in fields related to the life sciences and technologies.

In spite of the immense variety of organic semiconductors for typical electronic devices, there are generic features of the solid-state (crystalline) structures which we can identify and describe as representative. With that in mind, we have chosen a few examples, which we will describe in some detail.

7.2.1 A small molecule: pentacene

One of the most studied and used organic semiconductors, especially for organic field-effect transistors (OFETs), is pentacene. Its molecular, crystal, and electronic structures have been studied both in the bulk and in thin films grown on different substrates and under different conditions. In the Web of Science one can find more than 3000 references with 'pentacene' in the title. Its crystal and electronic structures were reported and discussed in references [24–38], among others. Its crystal structure is characterized by a herringbone arrangement with tilted edge-to-face π–π interactions, but the detailed structure of the bulk is different from those of the various polymorphs found in thin films, and there are at least four morphologies which can be identified by their $d(001)$ spacings, namely 14.1, 14.5, 15.0, and 15.4 Å.

Single crystals generally adopt the smallest $d(001)$ spacing of 14.1 Å in a triclinic $P\bar{1}$ structure with two molecules in the unit cell. Thin films grown on SiO_2 at temperatures of more than 350 K preferentially adopt the 14.5 Å $d(001)$ spacing and a slightly smaller unit cell volume [27]. Table 7.1 lists typical values of the unit cell dimensions for the structures of single crystals, thin films, and a monolayer on SiO_2.

The data set reported in the table for the thin film on SiO_2 is just one among many others, which for the *a*-axis, for example, range from 5.9 Å [32] to 6.485 Å [27], and different values are found for other substrates. In general, the thin-film unit cell is more rectangular than the unit cell in the bulk, and its two axes are shorter.

In OFETs, charge transport occurs in the first few monolayers, i.e. close to the dielectric. Therefore, the molecular packing in these first monolayers is particularly important, since it determines the transfer integrals—and the mobility—which depend on the overlap of the frontier orbitals in neighbouring molecules. Mannsfeld *et al* [34] found that in the first monolayer, pentacene molecules deposited on SiO_2 are oriented vertically with respect to the substrate when forming the usual herringbone pattern, in contrast with the common tilted arrangement observed in thin films and single crystals. Figure 7.3 illustrates the results they obtained from crystallographic refinement calculations in combination with grazing incidence x-ray diffraction (GIXD) data taken from a previous study of pentacene

Table 7.1. Crystal structures of pentacene at 293 K in the bulk (14.1 Å phase, single crystal), in a thin film (14.5 Å phase), and in a monolayer on a-SiO_2.

	a/Å	b/Å	c/Å	α/°	β/°	γ/°	V/Å^3	Ref.
Single crystal	6.266	7.775	14.530	76.475	87.682	84.684	685.15	[27]
Thin film	5.958	7.596	15.61	81.25	86.56	89.80	697	[31]
Monolayer	5.916	7.588				89.95		[29]

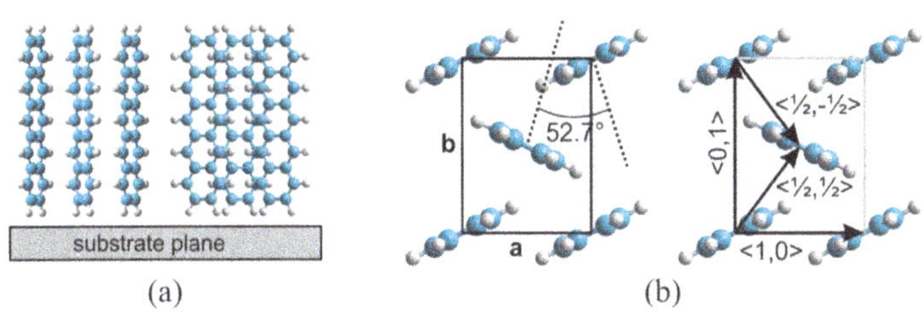

Figure 7.3. Alignment of pentacene molecules in the first monolayer. The two molecules in the unit cell are oriented vertically with respect to the substrate surface and form a herringbone pattern. (a) Single-cell side views: along the *a*-axis, left; along the *b*-axis, right. (b) *a*–*b* cell top view, left; *a*–*b* cell transfer integrals, right. Reproduced with permission from [34] and SSRL Science Highlight courtesy of the authors. Copyright Wiley-VCH Verlag GmbH and the authors.

monolayers [29]. In contrast to the vertical pentacene monolayer on SiO_2, a tilt of ca. 5 ° has been found in typical thin films on SiO_2 [31].

A quantum chemical calculation for the pentacene molecule, using the B3LYP/6-31G* density functional, gave the following values: $E_{HOMO} = -4.60$ eV, $E_{LUMO} = -2.39$ eV; these values are in agreement with calculated values found in the literature, for example, in [39], and close to the experimental values for sublimed commercial pentacene. In fact, some B3LYP density functionals give results which are in reasonable agreement with the experimental energy gaps.

The band structure of pentacene has also been studied using DFT in the bulk and in thin films, for example, those described in references [34, 37, 38 and 40], to cite but a few. There are two inequivalent molecules in the unit cell that lead to four bands near the Fermi level. Parisse *et al* [38], for example, found that the gap is smaller for the thin-film polymorph (0.57 eV for thin film vs 0.79 eV for the bulk) and reported the valence and conduction bandwidths to be ~0.7 and ~0.8 eV, respectively, which are larger than those of the bulk phase. All this is consistent with the higher conductivity observed in thin films. Mannsfeld *et al* [34] also calculated the transfer integrals for $t_{<1/2,\ 1/2>}$, which are shown in figure 7.3(b), and found values of 157 meV and 208 meV for the HOMO in the bulk and in the monolayer, respectively, and 215 meV and 207 meV for the LUMO in the bulk and in the monolayer, respectively. For $t_{<1/2,-1/2>}$, they reported 235 meV and 207 meV for the HOMO in the bulk and in the monolayer, respectively, and 192 meV and 207 meV for the LUMO in the bulk and in the monolayer.

7.2.2 Conjugated polymers

Polyacetylene is conceptually the simplest conjugated polymer and has been the subject of many expectations, particularly after the discovery in 1977 that its conductivity could be varied over several orders of magnitude by heavy doping and could reach up to 38 $S\,cm^{-1}$ at room temperature [13]. However, its potential uses in device applications have been hindered because it lacks processability, solubility, and stability. Its accurate crystal structure and detailed electronic structure have never been determined, in spite of much effort, and controversy still surrounds the question of whether it has bond alternation as a consequence of the Peierls instability, or whether it is just an agglomerate of finite chains and cross-linked polyene segments, whose conductivity is due to charge transfer between finite chains [21, 41, 42]. Nevertheless, the discovery of highly conductive doped polyacetylene was one of the most important milestones in the emergence of organic electronics, stimulating both theoretical and experimental research into organic semiconductors. The interesting physics of polyacetylene originates from its topological defects, as we will discuss below.

A few other conductive polymers have been known for a long time, such as doped polyaniline [43] and polypyrrole [44], and in fact, polyaniline exhibits true metallic transport [45]. These polymers are used in printed circuit board manufacturing, antistatic and electrostatic dissipative coatings, and corrosion protection.

P3HT is one of the most popular low-bandgap polymer semiconductors and is used in OSCs and OFETs due to its high charge-carrier mobility (up to 0.1 $cm^2 V^{-1} s^{-1}$). It is used at a wide range of molecular weights and regioregularities and has been a model polymer for research purposes. P3HT is just one of the many poly(3-alkylthiophenes) which have been synthesized and reported to undergo self-assembly and form oriented structures [46]. Thin films are usually only partially crystalline, which makes the determination of their structures very difficult, and several experimental methods, such as x-ray or electron diffraction, have had to be combined with quantum chemical calculations or molecular dynamics simulations. Several studies of the structure of P3HT have been reported, for example [46–51], and they are within reasonable agreement.

A good example is given by Dudenko *et al* who combined x-ray diffraction to assess long-range order in powder and fibres combined with high-resolution ^{1}H and ^{13}C solid-state magic angle spinning NMR spectroscopy, and unified the experimental results with quantum chemical and molecular dynamics calculations, a strategy they called 'multi-technique crystallography' [49]. They fully determined the bulk structure of the most stable phase of P3HT, which crystallizes in the monoclinic system and is in space group $P2_1/c$, with two non-equivalent polymer chains in the crystalline primitive cell. The results for the unit cell, obtained by deconvolution of the diffractograms into the individual contributions of the amorphous and crystalline parts, are reproduced in figure 7.4(b).

The electronic structure of regioregular P3HT has also been studied. Examples of reported values of the frontier levels for thin films are −5.12 eV for the HOMO and −2.84 eV for the LUMO [52]; and the direct energy gap at the Γ point is 0.78 eV for a single regioregular P3HT chain, as determined by DFT band structure calculations [48].

In organic electronics, different materials are chosen for different applications. For OFETs, charge-carrier mobility is one of the most important parameters taken

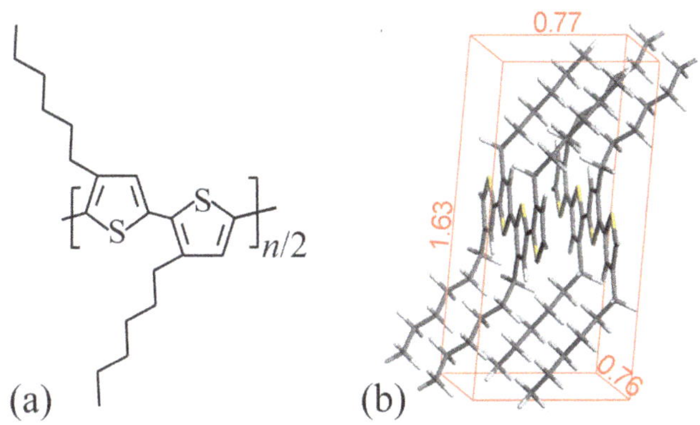

Figure 7.4. (a) Chemical structure of regioregular HT-P3HT. (b) Monoclinic unit cell ($\alpha = \beta = 90°$, $\gamma = 87°$) containing two non-equivalent chains with two monomers each. The numbers in red correspond to the cell dimensions in nanometers (regioregularity=94%). Reproduced with permission from [49], copyright Wiley-VCH Verlag GmbH.

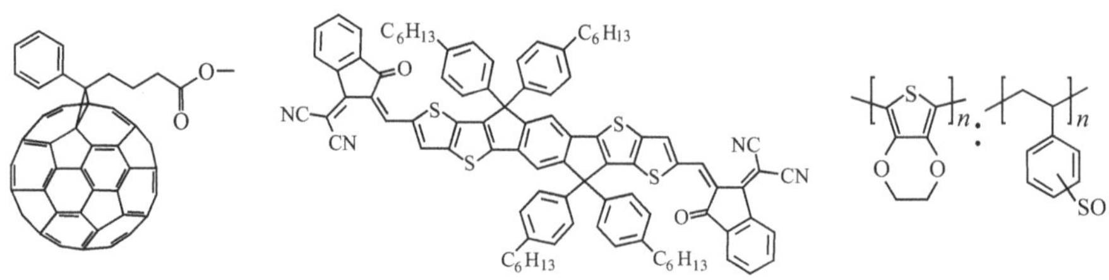

$PC_{61}BM$
Acceptor

ITIC
Acceptor

PEDOT:PSS

Figure 7.5. Electron acceptor $PC_{61}BM$ ($E_{HOMO} = -6.1$ eV; $E_{LUMO} = -3.7$ eV); ITIC (electron acceptor) $E_{HOMO} = -5.48$ eV, $E_{LUMO} = -3.83$ eV [53]; and PEDOT:PSS, which has metallic-type behaviour at polymer–metal interfaces with a work function of 4.7–5.2eV. Note that for P3HT, an electron-donor molecule, $E_{HOMO} = -5.12$ eV and $E_{LUMO} = -2.84$ eV.

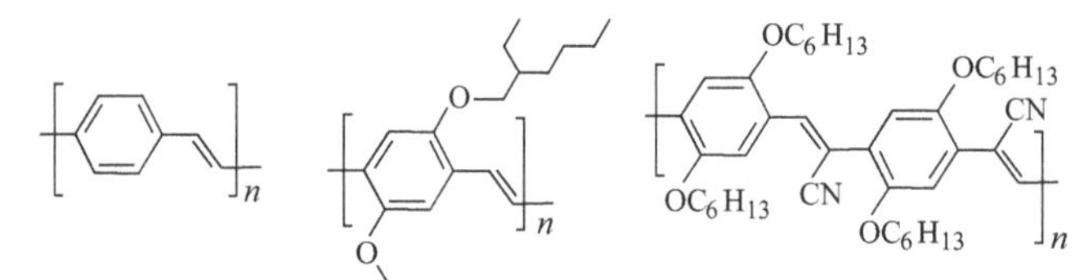

PPV

MEH-PPV

CN-PPV

Figure 7.6. PPV; MEH-PPV, its soluble derivative; and CN-PPV, a soluble PPV derivative with high electron affinity.

into account. Efficient OSCs require at least two kinds of materials, an electron donor and an electron acceptor, the latter of which is often a molecule, and many applications require high-conductivity materials. Figure 7.5 shows one of the most popular electron acceptors, PCBM, (a fullerene derivative), the non-fullerene acceptor known as ITIC, and PEDOT:PSS, which is one the most used conductive polymers due to its transparency and high work function.

7.2.3 From functionalization to molecular and crystal engineering

In typical organic electronic devices, most conjugated polymers behave as semiconductors. One advantage of organics is that the frontier levels, HOMO and LUMO, and as a consequence, the energy gap, can be chemically adjusted. A common procedure is functionalization, which consists of adding specific functional groups to give a compound new desired properties. A simple example is given in figure 7.6.

Figure 7.6 shows a starting material, PPV, which is insoluble (although it has a processable precursor), but its vinyl and phenyl groups can be functionalized to make a soluble derivative or modify its electronic properties. A very successful example is MEH-PPV, obtained by adding solubilizing alkoxy substituents to the arylene groups. This polymer has similar electronic properties and is soluble in common solvents such as chloroform and tetrahydrofuran. By introducing electron-withdrawing cyano groups at some of the ethylidene bonds and adding solubilizing long-chain alkoxy substituents to each arylene fragment, a soluble high-electron-affinity PPV derivative can be synthesized. The result is CN-PPV, which has lower HOMO and LUMO levels, and therefore an IP and an EA that are increased by ca. 0.5 eV, giving better electron injection and transport. Functionalization thus opened the way to improving or modifying the properties of organic electronic materials. An efficient two-layer PLED was obtained by combining CN-PPV with PPV [54], and an efficient OSC was fabricated using blends of MEH-PPV with CN-PPV [9]. These examples show how one can modify solubility properties and adjust the energies of the frontier levels by introducing small changes in the molecular structure. Note that small changes in the energies of the frontier levels can produce large changes in the behaviour of polymer/polymer interfaces [55].

As already mentioned, computational methods based on quantum chemistry or molecular dynamics can be of great help, both in determining the molecular and crystal structure as well as the morphology of materials, and by assisting in the design and synthesis of new materials. It is common practice among synthetic chemists working with organic electronics, especially those working with solar cells, to use computational quantum chemical methods to assist in the design of new materials. On the other hand, crystal engineering [56] is a highly sophisticated computational approach that can assist in the design and synthesis of organic crystalline solids by controlling their molecular structures, intermolecular interactions, packing arrangements, crystal growth, and patterning methods. Molecular self-assembly is at the core of crystal engineering, and consists of the spontaneous association of molecules, under equilibrial conditions, in order to form organized aggregates held together by non-covalent bonds (van der Waals interactions, hydrogen bonds, and the so-called π–π interactions), which are all, in fact, electrostatic interactions. Note that the generated structures must correspond to thermodynamic equilibrium and result from a favourable balance between the enthalpic and entropic terms in the Gibbs free-energy change ($\Delta G = \Delta H - T\Delta S \leqslant 0$).

7.3 Solitons, polarons, and bipolarons

Pure polyacetylene has room-temperature conductivity of the order of 10^{-5} S cm^{-1}, but it can be reduced to $<10^{-9}$ S cm^{-1} when exposed to NH_3 and increased to 220 S cm^{-1} when exposed to AsF_5, covering a range of eleven orders of magnitude [41].

The nature of the charge carriers in polyacetylene is different from that of the charge carriers in the other conjugated polymers. Polyacetylene has a degenerate ground state, i.e. two different geometries with the same energy, differing from each other by the exchange of the carbon–carbon single and double bonds (i.e. =C–C=C–··· and -C=C-C=···). In a polyacetylene chain there are bond-alternation defects, or misfits, where the chain breaks into domains of different geometry, corresponding to non-bonding $2p_z$ orbitals occupied by only one electron (see figure 7.7(a)). Each of these defects, which can move along the chain, corresponds to a bond-alternation parameter that changes sign at the misfit and behaves as a tidal wave, which is called a *soliton*.

A theory of solitons in conjugated polymers, developed by Su *et al* [57, 58], dealt with the creation energy, effective mass, extension, and transport activation energy;

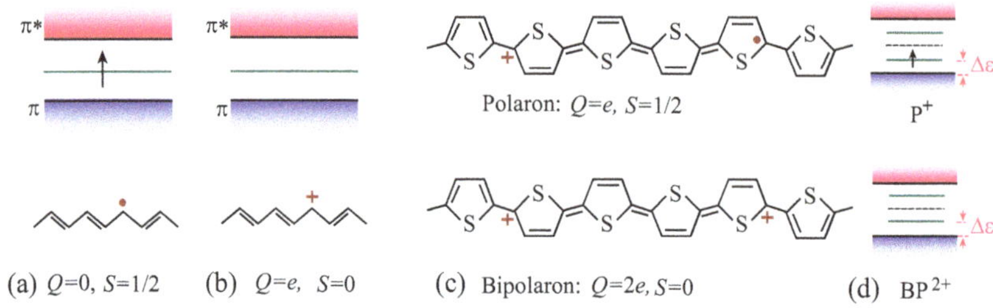

Figure 7.7. (a) A neutral soliton in polyacetylene: an unpaired electron on a non-bonding p_z orbital. (b) A positive soliton, the result of removing the electron from a neutral soliton. (c, d) The removal of one electron from a polythiophene chain causes a lattice distortion and shifts the HOMO level upward (and the LUMO downward) by $\Delta\varepsilon$. The distortion associated with the charge (due to the electron removal) forms a quasiparticle called a *polaron*. The removal of a second electron creates a *bipolaron*. (d) Polaron and bipolaron levels. The polaron binding energy is given by $E_\mathrm{p} = \Delta\varepsilon - E_\mathrm{distortion}$.

it explained the mobility of neutral solitons in undoped polyacetylene, as well as their role in the doping mechanism. The soliton energy levels are positioned in the middle of the gap between the π and π^* energy bands. They extend over several CH units and can move along the chain, providing charge transport with effective masses of the order of six times the free-electron mass. Neutral solitons (Figure 7.7(a)) have spin 1/2, but there are solitons with charges of $\pm e$ and spin zero (figure 7.7(b)). Doping can be used to create solitons with positive or negative charge and spin zero, neutral solitons with spin 1/2, and even solitons with fractional charges.

Most other conjugated polymers have non-degenerate ground states. In these polymers the addition or removal of one electron creates a bonding state and an anti-bonding state, which are accompanied by chain distortion. These charge carriers associated with lattice distortions are called *polarons* or *bipolarons* (see figures 7.7(c) and (d)). A bipolaron is a pair of charges associated with a large lattice distortion.

A description of solitons, polarons, and bipolarons in conjugated polymers was given by Brédas *et al* [59], and two good reviews are recommended, one on the theory of solitons by Heeger *et al* [60], and the other on polarons and excitons by Fisher *et al* [61].

7.4 Transport and optical properties

For applications in organic electronics we can consider two kinds of conjugated polymers: those that play the role of semiconductors in electronic devices and those that play the role of conductors in connecting lines and electrodes, such as PEDOT: PSS. The latter have high conductivity, similar to that observed in disordered metals. They show some metallic characteristics, such as Pauli-type magnetic susceptibility and linear temperature dependence of the thermoelectric power and specific heat, implying the existence of a nonzero density of states near the Fermi level. However, the dc conductivity is usually thermally activated, either decreasing with temperature or reaching a minimum below room temperature.

In a material in which there are different kinds of charge (e.g. electrons and holes in semiconductors), the conductivity is the sum of the contributions of all charge densities, n_i, $\sigma = \sum_i n_i q_i \mu_i$, where q_i is the charge (e.g. e) and μ_i is the mobility of charges of type i. μ_i can be written in terms of the mean scattering (or relaxation) time, τ_i, and the effective mass, m_i^*, as $\mu_i = e\tau_i/m_i^*$.

In typical inorganic semiconductors the conductivity is thermally activated and proportional to $\exp(-E_g/2k_BT)$, where E_g is the bandgap. In organic semiconductors, the charge carriers are generally polarons, and we can consider two different transport regimes, the band regime and the hopping regime. When the transfer integral between neighbouring molecular sites is large compared with other energies (due to disorder), the band model is valid. In this model, the mobility is related to the bandwidth and depends on the temperature according to $\mu \propto T^{-n}$, where $0 < n < 3$, due to electron–phonon and electron–electron interactions and impurities. Generally, band transport is possible if the mobility is greater than about 5 $cm^2\ V^{-1}\ s^{-1}$. Unfortunately, the room-temperature charge mobilities in organic semiconductors are usually low, in the range of 10^{-10}–10 $cm^2V^{-1}s^{-1}$, while in inorganic semiconductors they lie between 10^2 and 10^4 $cm^2\ V^{-1}\ s^{-1}$.

A transport theory of polaron hopping was developed by Holstein [62, 63] based on perturbation theory, which has been applied in the case of weak intermolecular interactions. He considered a boundary temperature, T_t, above which charge transport is due to random site jumps and below which a band description of polaron motion is valid. For $T > T_t$, Holstein derives a hopping rate which he introduces in the diffusion constant, D, and he gets the mobility through Einstein's equation $\mu_{hop} = eD/k_BT$. Here, we will not go into the long Holstein's arguments; we will go directly to the final results as described in the overview of charge transport in organic semiconductors given by Coropceanu *et al* [64], and in section 3.3. of the book by Köhler and Bässler, listed in Further reading.

Holstein's equation for the drift mobility at temperatures $T > T_t$, where the non-adiabatic small-polaron hopping regime is valid, is as follows:

$$\mu_{hop} = \frac{ea^2}{k_BT}k_{CT} \quad \text{with} \quad k_{CT} = \frac{2\pi}{\hbar}t^2\left(\frac{1}{4\pi\lambda k_BT}\right)^{1/2}\exp\left(-\frac{\lambda}{4\,k_BT}\right), \tag{7.1}$$

where a is the distance between sites, k_{CT} is the hopping rate (probability of charge transfer/hopping per unit time), λ is the reorganization energy induced by the charge transfer (which is about twice the binding energy (i.e. $\lambda \approx 2E_p$)), and t is the intermolecular transfer integral that accounts for the interaction between neighbouring molecules. The relations between the energy gap, the position of the polaron levels relative to the band edge, $\Delta\varepsilon$, the binding energy, E_p, and the reorganization energy, depend on many factors, such as the degree of doping. For polypyrrole (PPy), for example, the bandgap of the undoped PPy is 3.2 eV, E_p is 0.12 eV, and $\Delta\varepsilon$ is 0.49 eV for a low doping level, implying a reorganization energy, $E_{distortion}$, of about 0.37 eV [59]. This shows the complexity of the problem, and that equation (7.1) should be taken with a grain of salt.

To be practical, equation (7.1) predicts an Arrhenius-type temperature dependence:

$$\mu_{\text{hop}} \propto T^{-3/2} \exp(-E_a/k_B T),$$

where the activation energy is approximately half the binding energy. However, at high temperatures, when $E_p/2\,k_B T \ll 1$, the pre-exponential term $T^{-3/2}$ takes over. The crossover temperature between the two regimes is typically more than 600 K.

When the phonon coupling is weak, and at low temperatures, i.e. when $T < T_t$, ($T < 300$ K), the band model is valid, and the mobility follows T^{-n}. By increasing the temperature we may obtain a superposition of the decreasing band-type mobility and the increasing thermally activated hopping type, often going through a minimum before entering the $T^{-3/2}$ regime.

Another model for hopping transport is Mott's famous *variable-range hopping* (VRH), which is based on the assumption that transport occurs by hopping at the Fermi level, in which the charge carriers tunnel between localized states distributed over the whole gap [65]. Nevill F Mott shared the Nobel Prize in Physics 1977 with Philip W Anderson and John H van Vleck 'for their fundamental theoretical investigations of the electronic structure of magnetic and disordered systems'. At low temperatures, the phonon energies are low, favouring hopping over large distances. The conductivity is then given by the following equation:

$$\sigma(T) = \sigma_0 \exp\left[-(T_0/T)^{\gamma}\right]$$

where $\gamma = 1/(1 + D)$ is related to the dimensionality of the transport process ($D = 1$, 2, or 3). $T_0 \sim \beta/[k_B\, N(E_F)\zeta^3]$, where β is a numerical factor generally between 16 and 22, $N(E_F)$ is the density of localized states near the Fermi level, and ζ is the effective localization length. σ_0 is temperature dependent but is usually ignored, since the temperature dependence of the exponential term is dominant. $\gamma = 1/4$ for 3D hopping, but it is 1/2 for 1D hopping in disordered conducting polymers. PEDOT:PSS films have been described by this model [66] with $\gamma = 1/4$ (3D VRH).

The conductivity of transparent organic conductors such as PEDOT:PSS at terahertz (THz) frequencies shows a strong frequency dependence, allowing the determination of the plasma frequency and the scattering time. As we have discussed in 2.2, according to the Drude model, a conductor is transparent to electromagnetic radiation when the relative permittivity ε_r is positive and real, i.e. when the frequency is higher than the plasma frequency, ω_p (see equation 2.3). The Drude model is, however, an approximation (applicable to the electron gas), and more elaborate models have been developed, such as the Drude–Smith model [67]. This model can be used to model terahertz spectroscopy data, from which some important properties can be calculated, namely the terahertz conductivity, the charge-carrier density, and the mobility.

7.5 Spectroscopy and photophysics

The absorption and emission of radiation by molecules can be treated by time-dependent perturbation theory, which considers the energy of interaction of the electric field, $\mathbf{E}$, with the dipole moments, $\boldsymbol{\mu}$, of the molecules as a perturbation. To solve the problem we consider a time-dependent Schrödinger equation,

$$\mathrm{i}\hbar\frac{\mathrm{d}|\Psi(t)\rangle}{\mathrm{d}t}=[H_0+\lambda H'(t)]|\Psi(t)\rangle, \tag{7.2}$$

where H_0 is the unperturbed Hamiltonian for which the eigenvalues and eigenvectors are assumed to be known, $H_0|n\rangle = \varepsilon_n|n\rangle = \hbar\omega_n|n\rangle$, and $\lambda H'$ is the perturbation. Since the set $\{|n\rangle\}$ is complete, it can be used as a basis for the expansion of $|\Psi(t)\rangle$,

$$|\Psi(t)\rangle = \sum_n c_n(t)\exp(-\mathrm{i}\omega_n t)|n\rangle. \tag{7.3}$$

λ is assumed to be small which implies a weak time dependence of $c_n(t)$. By substituting (7.3) into (7.2), we can derive the rate of change of the coefficients:

$$\mathrm{d}c_m/\mathrm{d}t = -(\mathrm{i}/\hbar)\exp(\mathrm{i}\omega_{mn} t)\lambda H'_{mn}, \tag{7.4}$$

where $\omega_{mn} = \omega_m - \omega_n$, and $H'_{mn} = \langle m|H'|n\rangle$ are the matrix elements of the perturbation, i.e. the perturbation induces a transition between two stationary states $n \rightarrow m$ with energies ε_n and ε_m, such that $\varepsilon_m - \varepsilon_n = \hbar\omega_{mn}$. By convention, we use the indices '*mn*' in this sequence for any variable associated with the transition $n \rightarrow m$, where n and m are the initial and final states, respectively (e.g. $B_{n\rightarrow m} = B_{mn}$, $A_{m\rightarrow n} = A_{nm}$).

The perturbation is $H' = -\mathbf{E}\cdot\boldsymbol{\mu}$, where $\boldsymbol{\mu}$ is the dipole moment operator, with components $\mu_x = -ex$, etc. representing the dipole moment $\boldsymbol{\mu} = -\sum_i q_i\mathbf{r}_i$. The sum is over all the charged particles (nuclei and electrons) in the molecule, q_i is the charge on the ith particle, and $\mathbf{r}_i$ is its position. To make the calculation simple, we calculate using the x component and add the others at the end. The x component of an oscillating field can be written as

$$E_x = E_x^0\cos\omega t = (1/2)E_x^0[\exp(i\omega t) + \exp(-i\omega t)] \tag{7.5}$$

which appears in the perturbation, $H'_{mn} = -\langle m|E_x\mu_x|n\rangle$. By integrating (7.4) using (7.5) we get

$$c_m(t) = \frac{E_x^0}{2\hbar}\langle m|\mu_x|n\rangle\left\{\frac{\exp[\mathrm{i}(\omega_{mn}-\omega)t]-1}{\omega_{mn}-\omega}+\frac{\exp[\mathrm{i}(\omega_{mn}+\omega)t]-1}{\omega_{mn}+\omega}\right\}. \tag{7.6}$$

Equation (7.6) describes two types of process: the absorption of radiation associated with a transition from a state n to a state m of higher energy, $\omega_{mn} > 0$ (first term), and emission due to a transition from state n to a state m of lower energy, $\omega_{mn} < 0$, (second term). Note that the probability $|c_m|^2$ is appreciable only when one of the denominators is practically zero. As Einstein found out, emission can be spontaneous and stimulated [68].

Let us consider absorption first. It is relevant to relate the amplitude of the electric field, E_x^0, to the intensity of radiation. The intensity of radiation is the average value of the Poynting vector, $\mathbf{S} = \mathbf{E} \times \mathbf{H}$, which represents the flux of energy (energy transferred per unit area per unit time) in watts per square metre ($\mathrm{W\,m^{-2} = J\,s^{-1}m^{-2}}$).

In a propagating electromagnetic plane wave, the magnitudes of the magnetic field, $|\mathbf{H}|$, and the electric field, $|\mathbf{E}|$, are related by $|\mathbf{H}| = |\mathbf{E}|/\eta_0$, where $\eta_0 = \mu_0/\varepsilon_0$ is the intrinsic impedance of the vacuum. Since $1/\mu_0 = c\varepsilon_0^2$ (see note 2 in appendix C), we have $|\mathbf{H}| = c\varepsilon_0|\mathbf{E}| = c\varepsilon_0 E_x$. The average value of the Poynting vector is $\langle \mathbf{S} \rangle = (1/2)c\varepsilon_0(E_x^0)^2$, which is the intensity of the radiation, I, and is related to the energy density $\rho_x = I/c = (1/2)\varepsilon_0(E_x^0)^2$ (the cycle-averaged energy density in terms of E_x^0, in $\mathrm{J\,m^{-3}}$). This can be written in terms of the energy per unit volume for frequencies between ν and $\nu + \mathrm{d}\nu$ propagating in the x direction, $\rho_x = u_x(\nu)\mathrm{d}\nu$.

To calculate the transition probability, $P_{n\to m} = |c_m|^2$, we can use equation (7.6) and replace $(E_x^0)^2$ with a function of ρ_x. Noting that $|\exp(i\theta) - 1|^2 = 4\sin^2(\theta/2)$, we can write

$$|c_m(t)|^2 = \frac{2}{\varepsilon_0 \hbar^2}|\mu_{x,mn}|^2 \frac{\sin^2[(1/2)(\omega_{mn} - \omega)t]}{(\omega_{mn} - \omega)^2} \rho_x(\omega).$$

Up until now, we have considered monochromatic radiation with frequency ω. We now proceed to integrate over all frequencies, from 0 to ∞, or even from $-\infty$ to ∞, since the integral is only significant for values of the incident frequency near the transition frequency. We can then use $\omega = 2\pi\nu$, and, since the radiation is isotropic, $u_x(\nu_{mn}) = u_y(\nu_{mn}) = u_z(\nu_{mn}) = (1/3)u(\nu_{mn})$, we get, for a randomly oriented sample for which $\left| \mu_{x,mn} \right|^2 = (1/3)|\boldsymbol{\mu}_{mn}|^2$,

$$|c_m(t)|^2 = \frac{1}{6\hbar^2\varepsilon_0} |\boldsymbol{\mu}_{mn}|^2 u(\nu_{mn})\, t.$$

We can now derive the transition rate. We will use ε in place of ε_0 since the processes happen in a material, and recall that $v = c/n$ for the phase speed of light in a medium of refractive index n, noting that $\varepsilon_\mathrm{r} = \varepsilon/\varepsilon_0 = n^2$:

$$\frac{\mathrm{d}P_{n\to m}}{\mathrm{d}t} = \frac{\mathrm{d}\,|c_m(t)|^2}{\mathrm{d}t} = \frac{1}{6\hbar^2\varepsilon} |\boldsymbol{\mu}_{mn}|^2 u(\nu_{mn}) = B_{mn}^{\nu}\, u(\nu_{mn}),$$

where B_{mn}^{ν} is Einstein's absorption coefficient:

$$B_{mn}^{\nu} = \frac{1}{6\hbar^2\varepsilon} |\boldsymbol{\mu}_{mn}|^2, \quad \text{or} \quad B_{mn}^{\omega} = \frac{\pi}{3\hbar^2\varepsilon} |\boldsymbol{\mu}_{mn}|^2, \qquad \text{with } [B_{mn}^{\nu,\omega}] = \mathrm{J^{-1}\,m^3\,s^{-2}}.$$

If we use $u(\nu)$ (energy density per unit frequency interval, i.e. $\mathrm{J\,m^{-3}Hz^{-1}}$), then, since $B_{mn}^{\omega}u(\omega) = B_{mn}^{\nu}u(\nu)$ and $u(\omega)\mathrm{d}\omega = u(\nu)\mathrm{d}\nu$, we have $B_{mn}^{\nu} = B_{mn}^{\omega}/2\pi$ when using the angular frequency, ω.

The easiest way to get the spontaneous and stimulated emission coefficients, A_{nm} and B_{nm}, respectively, is to use Einstein's arguments and his A/B ratio [68], which he found by making $u(\nu_{mn})$ agree with Planck's law for black body radiation. He found that $B_{nm} = B_{mn}$ and $A_{nm}/B_{mn}^{\nu} = 8\pi h\nu_{mn}^3/c^3$, from which we can derive A_{nm}:

$$A_{nm} = \frac{8\pi^2\nu_{mn}^3}{3\hbar\varepsilon c^3}|\boldsymbol{\mu}_{mn}|^2 = \frac{\omega^3}{3\pi\hbar\varepsilon c^3}|\boldsymbol{\mu}_{mn}|^2, \qquad \text{with } [A_{nm}] = \mathrm{s}^{-1}.$$

Einstein's coefficients give the rates of the transitions between a state n and a state m with higher energy. Note that we can immediately define the radiative lifetime as $\tau_{\mathrm{rad}} = 1/A_{nm}$.

An electronic transition can be characterized in terms of the (unitless) oscillator strength, f, which is defined in terms of B_{mn}, or A_{nm}, or the transition dipole moment, as

$$f_{mn} = \frac{4\varepsilon m_e \hbar\omega_{mn}}{e^2}B_{mn}^{\nu} = \frac{2\pi\varepsilon m_e c^3}{\omega_{mn}^2 e^2}A_{mn} = \frac{2m_e\omega_{mn}}{3\hbar e^2}|\boldsymbol{\mu}_{mn}|^2$$

where m_e and e are the electron rest mass and the elementary charge, respectively.

The oscillator strength can also be derived in quantum mechanics terms by applying the Thomas–Reiche–Kuhn sum rule, which states that the sum of the probabilities for an electron to undergo a transition from its ground state into all excited states is one, i.e. $\sum_{m \neq n} f_{mn} = 1$.

In optical spectroscopy the absorption of radiation is usually expressed in terms of several quantities, namely the *absorbance*, or *optical density*, A; the *absorption coefficient*, α; the *absorption cross section*, σ; and the *molar extinction coefficient* or *molar absorptivity*, ε_A, given by the Beer–Lambert Law expressed in different ways:

$$I = I_0\, \mathrm{e}^{-\alpha C\ell} = I_0\, \mathrm{e}^{-\sigma N\ell} = I_0\, 10^{-\varepsilon_A C\ell} \equiv I_0\, 10^{-A}$$

where I_0 is the incident light intensity, I is the light intensity after passing through a sample of path length ℓ (in cm), and C is the concentration (in moles per litre), $[C] = \mathrm{mol\ L^{-1}}$. Note that A is dimensionless, and $A = \varepsilon_A C\ell$, implying that ε_A has units of $[\varepsilon_A] = \mathrm{L\ mol^{-1}\ cm^{-1}}$. The absorption cross section has units of $[\sigma] = \mathrm{cm}^2$, and N is the number density in units $[N] = \mathrm{molecules\ cm^{-3}}$; α has units of $[\alpha] = \mathrm{L\ mol^{-1} cm^{-1}}$.

It is important to note that the *integrated absorption coefficient* (the area under the absorption curve) is related to the oscillator strength:

$$S = \int_{\nu_1}^{\nu_2} \alpha(\nu)\, \mathrm{d}\nu = \frac{N_A\, e^2}{4m_e c\varepsilon} f_{mn},$$

where N_A is Avogadro's number and e, m_e, c and ε are the usual quantities.

Absorption spectra are often expressed as wavelengths, i.e. $[\lambda] = \mathrm{nm}$, Å, etc. However, other units are used, such as frequency $[\nu = c/\lambda] = \mathrm{s}^{-1}$, angular frequency $[\omega = 2\pi\nu = 2\pi c/\lambda] = \mathrm{rad\ s^{-1}}$, wavenumbers $[\bar{\nu} = \nu/c = 1/\lambda] = \mathrm{cm}^{-1}$, energy $[E = h\nu = hc/\lambda = \hbar\omega] = \mathrm{J}$ (joules), eV (electronvolts), kcal mol^{-1}, etc.; 1 eV $= 1.6 \times 10^{-19}$ J $= 23.06$ kcal mol^{-1}.

One can often observe a vibrational structure in the optical spectra, due to the fact that the vibrational energies are non-negligible when compared to the energies of the electronic transitions. It is then appropriate to consider the molecular wave function as the product of an electronic component and a vibrational component. Within the adiabatic Born–Oppenheimer approximation, which considers that in a molecule the wave functions of electrons and nuclei can be treated separately, the wave function can be written as

$$\Psi(\mathbf{r}, \mathbf{R}) = \psi_e(\mathbf{r}, \mathbf{R})\, \chi_v^e(\mathbf{R})$$

where **r** are the electron coordinates and **R** are the nuclear coordinates. The electronic wave function, $\psi_e(\mathbf{r}, \mathbf{R})$, depends on both the electron and the nuclear coordinates, and $\chi_v^e(\mathbf{R})$ is the wave function which describes the motion of the nuclei in the ψ_e electronic state. The index e corresponds to an electronic state, while v is a vibrational state. According to the Condon approximation, the transition dipole moment for a transition from the electronic state n in the vibrational state v to the electronic state m and vibrational state v' is $\boldsymbol{\mu}_{mv',nv} = \boldsymbol{\mu}_{mn}^0 \langle \chi_{v'}^m | \chi_v^n \rangle$, where $\boldsymbol{\mu}_{mn}^0$ is the first term of a series expansion of $\boldsymbol{\mu}_{mn}$. The intensity (e.g. absorption cross section) of the vibrational transitions is proportional to the square of the overlap integral between the vibrational wave functions involved in the transition, i.e. $\sigma \propto |\boldsymbol{\mu}_{mv',nv}|^2 = |\boldsymbol{\mu}_{mn}^0|^2 \times FC$, where FC is the Franck–Condon factor $|\langle \chi_{v'}^m | \chi_v^n \rangle|^2$.

It may be opportune to remind ourselves of the importance of the spin multiplicity in spectroscopy and photophysics. Since the full electron wave functions must be antisymmetric, there are four possibilities. If the spatial wave function is symmetric, the spin component must be antisymmetric, and we have a singlet state with $S = 0$ and $M_s = 0$ (see schematic on the left, below). If the spatial wave function is antisymmetric, the spin component must be symmetric, and we have a triplet state with $S = 1$ and $M_s = 1, 0, -1$ (schematic on the right). This is important for OLEDs because, statistically speaking, only 25% of the electrically generated excited states are singlets, while 75% are triplets:

$$\frac{1}{\sqrt{2}}(|\uparrow\downarrow\rangle - |\downarrow\uparrow\rangle),\ S = 0,\ M_s = 0 \qquad \begin{array}{ll} |\uparrow\uparrow\rangle, & S = 1,\ M_s = 1 \\ \frac{1}{\sqrt{2}}(|\uparrow\downarrow\rangle + |\downarrow\uparrow\rangle), & S = 1,\ M_s = 0 \\ |\downarrow\downarrow\rangle, & S = 1,\ M_s = -1 \end{array}$$

When a molecular system absorbs a photon, undergoing a transition from a singlet ground state S_0 to the vibrational manifold of an electronic excited state S_n, several processes can occur, including non-radiative and radiative transitions which include fluorescence and phosphorescence. Figure 7.8 schematically shows the most relevant electronic processes and transitions, in a Jablonski diagram [69].

Absorption occurs on a time scale of 10^{-15} s (femtoseconds), and is followed by vibration relaxation or internal conversion (IC) to the vibrational ground state of the excited singlet S_1, in about 10^{-12} to 10^{-10} s (picoseconds). The lifetime of the excited singlet is of the order of 10^{-10} to 10^{-7} s (nanoseconds) and the subsequent emission of a photon corresponds to fluorescence, F, (with a longer wavelength than that of the

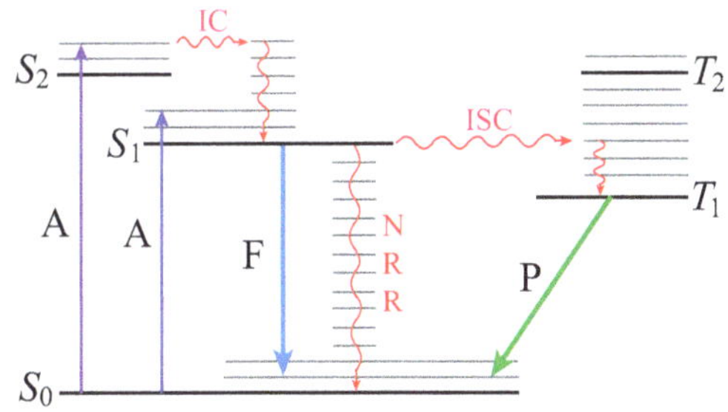

Figure 7.8. Jablonski diagram for the main possible processes involving the singlet (S_n) and triplet (T_n) manifolds. A: Absorption, F: fluorescence, P: phosphorescence, IC: internal conversion, ISC: intersystem crossing, NRR: non-radiative relaxation.

absorbed photon). In molecules with a high degree of spin–orbit coupling, intersystem crossing (ISC) from the singlet to the triplet manifold may occur and lead to a ('forbidden') transition from the triplet state to the singlet ground state. The intersystem crossing occurs in ca. 10^{-10} to 10^{-8} s (ps to ns) and the lifetime of the triplet excited state is of order of 10^{-6} to 1 s; the ensuing emission corresponds to phosphorescence. Vibrational non-radiative relaxation (NRR) processes of the redistribution of vibrational energy from optically coupled modes to 'dark' modes may occur as well.

7.5.1 Excitons

Excitons are electron–hole, (*e–h*), pair bound states (due to the Coulomb interaction) that result from the excitation of electrons from the valence band (or HOMO energy level), leaving holes behind. The electron and hole may have antiparallel spins ($S = 0$) or parallel spins ($S = 1$), i.e. they are either *singlet* or *triplet* excitons. As they have integer spin, they are quasi bosons (not real bosons). They are elementary excitations of condensed matter that can transport energy but not electric charge. They are monoenergetic and can be described by Bloch functions, which implies that their energy levels can form exciton bands. In real life, excitons appear in superpositions, as wave packets, with finite energy bandwidths, and transport energy according to their group velocities and effective masses. In both inorganic and organic semiconductors, excitons are created as the result of photon absorption. An exciton has a crystal momentum $\mathbf{K} = \mathbf{k}_e - \mathbf{k}_h$ and a spin $\sigma = s_e - s_h$. To obtain the energy eigenvalues in the exciton representation, a change of basis has to be performed from the separate electron n and hole m states by using linear combinations of the Slater determinants $|m\mathbf{k}\rangle$ and diagonalizing the energy matrix in the new basis, $|mn\mathbf{K}\rangle$. Since the exciton wave functions must satisfy the Bloch theorem, they are of the form given by equation (4.18), i.e.

$$\psi_{mn\mathbf{K}}(\mathbf{R}) = u_{mn\mathbf{K}}(\mathbf{R})\, e^{i\,\mathbf{K}\cdot\mathbf{R}} \tag{7.7}$$

where $\mathbf{R}$ is the position of the center of mass of the *e–h* pair. Figure 7.9 represents, in a 2D lattice (for clarity), the weakly bound Wannier–Mott exciton typical of

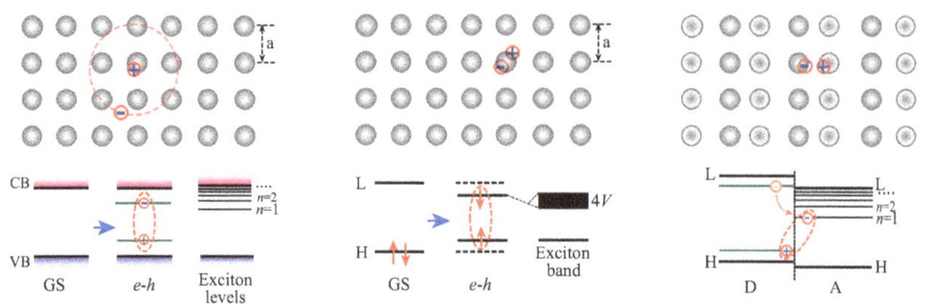

(a) Wannier-Mott exciton (b) Frenkel exciton (c) Charge transfer exciton

Figure 7.9. (a) The Wannier–Mott exciton. VB and CB stand for the valence and conduction bands, respectively, and GS is the ground state before the creation of the exciton. (b) The Frenkel exciton. H and L are the HOMO and LUMO levels, respectively. $4V$ is the exciton bandwidth. (c) The charge-transfer (CT) exciton at a donor/acceptor (D/A) interface: an electron is transferred from a donor molecule to an exciton level of a neighbouring acceptor molecule. H and L are the HOMO and LUMO levels, respectively.

inorganic semiconductors (a), the tightly bound Frenkel exciton (b), and the charge-transfer exciton at a donor/acceptor interface (c) typical of organic materials.

Typical values of the radius and binding energy for the Wannier–Mott exciton are ~10 nm and ~10 meV, respectively, while the corresponding values for the Frenkel exciton are ~1 nm and ~1 eV. The Wannier–Mott exciton is a weakly bound and delocalized *e–h* pair (figure 7.9(a)). Its motion is described by a Bloch function such as (7.7), which is the product of a plane wave and a hydrogen-like wave function which is the solution of a Schrödinger equation with a Coulomb potential $V(\mathbf{r}) = -e^2/4\pi\varepsilon r$, where $r = |\mathbf{r}_\mathrm{h} - \mathbf{r}_\mathrm{e}|$ is the distance between the electron and the hole.

The energy in the band picture is the energy of the gap, E_g, minus the binding energy, E_b. The latter is similar to the energy of the hydrogen atom plus the kinetic energy of the centre of mass of the exciton:

$$E_n(\mathbf{K}) = E_\mathrm{g} - \frac{R}{n^2} + \frac{\hbar^2 K^2}{2(m_\mathrm{e}^* + m_\mathrm{h}^*)}, \quad \text{with} \quad R = R_\mathrm{H}\frac{\mu_\mathrm{exc}}{\mu_\mathrm{H}}\frac{1}{\kappa^2}, \tag{7.8}$$

where $-R/n^2$ (with $n = 1, 2, \ldots$) are the energy eigenvalues of the bound *e–h* pair (i.e. the exciton), which are located below the conduction band minimum (figure 7.9(a)). R is the adapted Rydberg constant, with $R_\mathrm{H} = \mu_\mathrm{H} e^4/(32\pi^2\,\varepsilon^2\,\hbar^2) = 13.6$ eV ($n = 1$ energy level for the H atom), $\mu_\mathrm{H}/\mu_\mathrm{exc}$ is the ratio between the reduced masses of the H atom and the exciton, where $\mu_\mathrm{H} \approx m_\mathrm{e}$, and $\mu_\mathrm{exc} = m_\mathrm{e}^* m_\mathrm{h}^*/(m_\mathrm{e}^* + m_\mathrm{h}^*)$, and $\kappa = \varepsilon/\varepsilon_0$ is an *effective dielectric constant.*

The wave functions (7.7) are identical to those of the H atom multiplied by a plane wave, $\Psi \sim F_{n\ell m}(\mathbf{r})\exp(\mathrm{i}\,\mathbf{K}\cdot\mathbf{R})$. For example, for $n = 1$, $F_{1s}(\mathbf{r}) = (1/\sqrt{\pi\, a_\mathrm{exc}^3})\exp(-r/a_\mathrm{exc})$, with $a_\mathrm{exc} = \kappa\, m_\mathrm{e} a_0/\mu_\mathrm{exc}$, where $a_0 = 0.529$ Å, the Bohr radius. The exciton radius for the nth state is given by $r_n = a_\mathrm{exc} n^2$.

When the kinetic energy distribution of the excitons does not correspond to the equilibrial energy distribution at the crystal temperature, there may be excitons whose kinetic energy exceeds the mean thermal energy k_BT. These excitons are called *hot excitons*, and they have a high probability of dissociation or of being trapped by defects [70].

In organic molecular crystals (e.g. pentacene) the low dielectric constant (typically in the range 2–4) implies a large Coulomb interaction between the electron and the hole, leading to Frenkel-type excitons which are strongly bound *e–h* pairs, with the electron and the hole at the same lattice site (figure 7.9(b)). The binding energies are of the order of ~0.5–1.5 eV. In a crystal, the exciton motion can be described by a Bloch function in a tight binding, first-neighbours approximation. The energy eigenvalues, $E(\mathbf{K})$, are given by an equation like (5.1):

$$E(K) = E_{\text{exc}} + 2\,V\cos(Ka), \qquad \text{with} \quad K = 2\pi m/Na; \quad m = -N/2, \ldots, 0, \ldots, N/2,$$

where E_{exc} is the exciton energy in the molecule, V is the energy associated with the motion of the exciton from a molecule to one of its first neighbours (at distance a), and N is the number of lattice sites. An exciton bandwidth can be defined as $4V$.

The fact that photons have a small wave vector implies that $Ka \ll 1$, to satisfy the conservation of momentum in the creation of excitons. Given this condition, we can estimate the effective mass of the exciton, M^*_{exc}, in terms of the bandwidth by expanding the cosine in a Maclaurin series and taking only the first terms, i.e. $\cos(Ka) \cong 1 - K^2a^2/2$, and get:

$$E(K) \approx E_{\text{exc}} + 2V - VK^2a^2 = E_{\text{exc}} + 2V + \frac{\hbar^2K^2}{2\,M^*_{\text{exc}}},$$

where it is assumed that $-VK^2a^2$ corresponds to the kinetic energy, $\hbar^2K^2/(2\,M^*_{\text{exc}})$. Therefore,

$$M^*_{\text{exc}} = \hbar^2\left[\frac{\partial^2E(K)}{\partial K^2}\right]^{-1} = -\frac{\hbar^2}{2Va^2}.$$

Although the energy band model can be applied to Frenkel excitons (see figure 7.9(b)), these excitons are molecular excited states and it is appropriate to treat them as such: exciton creation consists in an electronic transition from the HOMO level to the LUMO level, followed by an upward displacement of the HOMO level and a downward displacement of the LUMO level, due to the Coulomb interaction between the electron and the hole, this being responsible for the binding energy.

At a donor/acceptor (D/A) interface (e.g. in organic solar cells), the exciton in the donor can dissociate into a charge-transfer (CT) exciton by electron transfer from the donor molecule to an exciton level of a neighbouring acceptor molecule (figure 7.9(c)). The CT exciton can be considered to be an intermediate between the Wannier–Mott and the Frenkel excitons and, as in the former, its energy levels are similar to those of the H atom, with principal and angular momentum quantum numbers. The binding energy of the lowest $\text{CT}_{1\text{s}}$ exciton is typically higher than k_BT at room temperature, and therefore has a very low probability of dissociation. The fact that CT excitons play an important role in organic heterojunction solar cells, i.e.

that their charges do have a high probability of separating, can be explained by the assumption that the CT excitons are hot excitons formed across the D/A interface and have low binding energy [71].

7.5.1.1 Exciton binding energy

The exciton binding energy of inorganic semiconductors is $E_b = E_g - E_a$, where E_g is the bandgap and E_a is the optical absorption edge (exciton creation energy). In organic semiconductors, the excitons are, in fact, *polaron–excitons* [72]—when electrons and holes are injected into the material, they go into polaron levels P^- and P^+ in the gap. The organics' analogue of the inorganic semiconductor bandgap is called the *single-particle energy gap*, E_{sg}, and is the difference between the electron and the hole injection energy levels (the sum of the electron and hole Schottky barriers of the metal contact) [73–75]. For E_a Campbell *et al* [73] took the first peak in the absorption edge ($0 \rightarrow 0$ transition). They used $E_{sg} = 2.45$ eV and $E_a = 2.25$ eV to get the exciton binding energy $E_b = 0.2$ eV for MEH-PPV.

The exciton binding energy is usually assumed to be due to the *e–h* Coulomb interaction. This is reasonable because the contributions to the binding energy from electron correlation and electron–lattice coupling practically cancel out [75]. Liu *et al* [76], for example, used a simplified wave function for a particle in a three-dimensional box, ψ_{n_x,n_y,n_z}, where n_x, n_y, and n_z are the quantum numbers, and the box dimensions are chosen according to the molecular dimensions. The exciton binding energy is given by

$$E_b = \frac{e^2}{4\pi\kappa\varepsilon_0} \iint \frac{|\psi_{0,0,0}(\mathbf{r}_h)|^2 \, |\psi_{1,0,0}(\mathbf{r}_e)|^2}{|\mathbf{r}_h - \mathbf{r}_e|} d\mathbf{r}_h \, d\mathbf{r}_e,$$

where $\psi_{0,0,0}(\mathbf{r}_h)$ and $\psi_{1,0,0}(\mathbf{r}_e)$ are the hole and electron wave functions, respectively, κ is an effective relative electric permittivity that accounts for the dielectric constant of the molecule and the relaxation energy of the ionic species, and $|\mathbf{r}_h - \mathbf{r}_e|$ is the size of the exciton. Remarkably, a value of $\kappa = 5.4$ approximately fits a whole series of acceptors, ranging from fullerenes to non-fullerenes. More sophisticated calculations have also been done by Liu *et al* using density functional theory and quantum mechanical modelling.

7.5.1.2 Excited state energy transfer

The wave-like propagation of excitons is not the only type of energy propagation in a molecular system. Energy transfer can occur via many processes, in which the coherent nature of the transport process is lost and energy 'jumps' from one site to another. Three main processes are generally considered, depending on the distance over which they occur: (i) long-range radiative, or trivial, energy transfer, (ii) Förster resonant energy transfer (FRET), and (iii) Dexter, or exchange, energy transfer. The total rate of energy transfer is the sum of all active processes.

Radiative energy transfer involves the emission of a photon by an excited donor, D*, followed by absorption by an acceptor, A, a considerable distance away ('*' denotes the excited states):

$$D^* \rightarrow D + h\nu$$
$$h\nu + A \rightarrow A^*.$$

The transfer rate depends on the distance between donor and acceptor, R_{DA}, as $k_{ET} \propto 1/R_{DA}^2$ and is typically of the order of 10–50 nm when neither exchange transfer nor FRET are active.

The relevant cases of donor–acceptor energy transfer are, in principle, one-step non-radiative processes driven by quantum coupling (QC):

$$D^*+A \xrightarrow{QC} D + A^*$$

Quantum coupling involves electron–electron interactions which, as we know from quantum mechanics, have two components: the Coulomb interaction and the exchange interaction, $V_{ee} = V_C - V_{Exc}$. As discussed in chapter 3, in a multi-electron system such as a donor–acceptor molecular pair, it is convenient to consider the interaction between charge distributions using the multipole expansion of the potential.

The rate constant for energy transfer between a donor and an acceptor molecule can be treated by perturbation theory. The following equation is obtained:

$$K_{ET} = \frac{2\pi}{\hbar} V_{ee}^2 \rho_E, \qquad \text{with} \quad V_{ee} = \langle DA^* | H' | D^* A \rangle,$$

where $|DA^*\rangle$ and $|D^*A\rangle$ are the Slater determinants for the final and initial states, in terms of the donor, $|D\rangle$, and acceptor, $|A\rangle$, states, respectively. H' is the perturbation Hamiltonian describing the electrostatic interaction between the charge distributions in the donor and acceptor, and ρ_E is the density of states related to the spectral overlap, J, between the donor and acceptor emission spectra.

In this context it is instructive to separate the electrostatic interaction into the Coulomb, V_C, and the exchange, V_{Exc}, terms, which we can do by also considering that the wave function has a spatial component, ψ_r, and a spin component, ψ_σ, i.e. $\Psi = \psi_r \psi_\sigma$. We can then write, for V_{ee}:

$$V_{ee} = V_C - V_{Exc}$$

$$V_C = \langle D_r(1) A_r^*(2) |\, H' \,|\, D_r^*(1) A_r(2) \rangle \langle D_\sigma(1) |\, D_\sigma^*(1) \rangle \langle A_\sigma^*(2) |\, A_\sigma(2) \rangle \tag{7.9}$$

$$V_{Exc} = \langle D_r(1) A_r^*(2) |\, H' \,|\, D_r^*(2) A_r(1) \rangle \langle D_\sigma(1) |\, A_\sigma(1) \rangle \langle A_\sigma^*(2) |\, D_\sigma^*(2) \rangle, \tag{7.10}$$

where the indices 1 and 2 represent two indiscernible interacting electrons, one in the donor, the other in the acceptor.

We may now consider the two extreme mechanisms of energy transfer between donor and acceptor molecules: (i) if the separation between the donor and acceptor molecules is large compared with the molecular size, the dominant coupling is a dipole–dipole Coulomb interaction (through space), and the process is referred as *Förster transfer* [77]; (ii) if the excited donor molecule is in contact with the acceptor, the quantum coupling is primarily due to the exchange interaction, and the process is known as *Dexter transfer* [78].

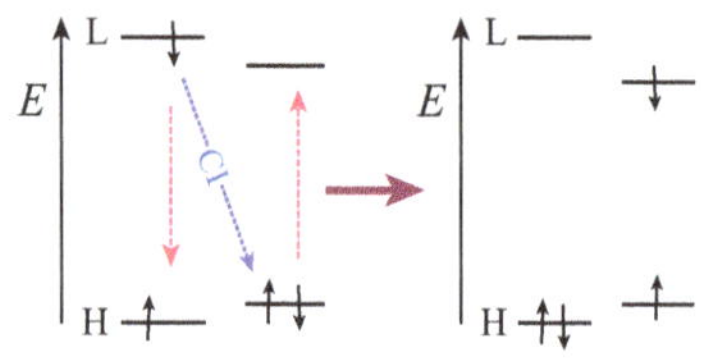

Figure 7.10. Energy level diagram for a Förster energy transfer. H and L stand for the HOMO and the LUMO, respectively, and CI stands for the Coulomb interaction.

The Coulomb term (7.9) is only nonzero if the ground and excited states of the donor have the same spin multiplicity, the same being true for the acceptor; but the multiplicities of the donor and acceptor can be different. For this reason, Förster transfer can occur between a donor in a singlet excited state and an acceptor in a singlet ground state, or, in principle, between a donor in a triplet T_n state and an acceptor in a triplet T_1 state. The Förster mechanism is schematically represented in figure 7.10 for an energy transfer via the Coulomb interaction from a donor molecule in a singlet excited state to an acceptor molecule in the singlet ground state which becomes excited. The selection rule $\Delta S = 0$, which requires that the initial and final spin states of D, and A must be identical, may be relaxed by spin–orbit coupling effects when D contains a heavy metal atom. In this case the process described by the equation in figure 7.10 is converted into:

$$^3D^{*} + {}^1A \rightarrow {}^1D + {}^1A^{*}.$$

This process is known as phosphor-sensitized fluorescence and is a way to obtain highly efficient fluorescent OLEDs [79, 80].

In the Förster energy transfer, the Coulomb term is predominant and can be expressed as a function of the transition dipole moments of the donor and acceptor molecules according to $V_{\mathrm{C}} \equiv V_{\mathrm{dd}} \propto \boldsymbol{\mu}_{\mathrm{D}} \cdot \boldsymbol{\mu}_{\mathrm{A}} / R_{\mathrm{DA}}^3$, where R_{DA} is the D–A distance (see equation 3.11 for an interaction between two polar molecules at a fixed orientation). The rate depends on V_{C}^2, implying that $k_{\mathrm{ET,C}} \propto 1/R_{\mathrm{DA}}^6$. Thus, we can write

$$k_{\mathrm{ET,C}} \propto \frac{f_{\mathrm{D}} f_{\mathrm{A}}}{R_{\mathrm{DA}}^6} J,$$

where f_{D} and f_{A} are the oscillator strengths of the donor and acceptor, respectively. The spectral overlap integral J is given by

$$J(\lambda) = \int I_{\mathrm{D}}(\lambda)\, \lambda^4\, \varepsilon_{\mathrm{A}}(\lambda)\, \mathrm{d}\lambda$$

where I_{D} is the wavelength-dependent donor emission (fluorescence) spectrum normalized to an area of one, $\int I_{\mathrm{D}}(\lambda)\mathrm{d}\lambda = 1$, and ε_{A} is the molar extinction coefficient spectrum of the acceptor. ε_{A} is expressed in $\mathrm{cm}^{-1}\,\mathrm{mol}^{-1}\,\mathrm{L}$, λ in nm, and J is in units of $\mathrm{cm}^{-1}\,\mathrm{mol}^{-1}\,\mathrm{L}\,\mathrm{nm}^4$.

Förster derived the following equation for the rate of energy transfer using spectroscopic data [77] (see Köhler and Bässler's section 2.2.4 and Forrest's section 3.8 in Further reading for details):

$$k_{\mathrm{ET,C}} = \frac{1}{\tau_{\mathrm{D}}}\left(\frac{R_0}{R_{\mathrm{DA}}}\right)^6, \quad \text{with} \quad R_0 = \left[\frac{9\,(\ln 10)\,\kappa_{\mathrm{F}}^2 \Phi_{\mathrm{D}}}{2^7 \pi^5 N_{\mathrm{A}} n^4} J(\lambda)\right]^{1/6}, \tag{7.11}$$

where τ_{D} is the radiative lifetime of the donor in the absence of the acceptor, R_0 (in nm) is the Förster radius or transfer distance, and Φ_{D} is the fluorescence quantum yield of the donor in the absence of the acceptor. $\kappa_{\mathrm{F}} = \cos\theta_{\mathrm{DA}} - 3\cos\theta_{\mathrm{D}}\cos\theta_{\mathrm{A}}$ is the orientation factor of equation (3.11), and n is the refractive index of the medium in which the donor and acceptor are immersed. The efficiency of the energy transfer may be defined as

$$\eta_{\mathrm{ET,C}} = \frac{k_{\mathrm{ET,C}}}{(1/\tau_{\mathrm{D}}) + k_{\mathrm{ET,C}}} = \frac{1}{1 + (R_{\mathrm{DA}}/R_0)^6}.$$

The typical transfer distance is generally <10 nm, with typical values in range of 1–4 nm.

The ideal Förster mechanism, as presented here, implies that donor–acceptor energy transfer is a one-step process, but in real life, it is often a multistep and complex process.

The Dexter mechanism [78] involves electron exchange between a donor and an acceptor (equation 7.10). When the excited donor is in contact (<1 nm) with the acceptor, an electron from the LUMO of the donor is transferred to the LUMO of the acceptor simultaneously with the transfer of an electron from the acceptor's HOMO to the donor's HOMO. This mechanism requires the overlap of the donor and acceptor wave functions in order to enable electron exchange between the respective orbitals. The exchange can occur between states of the same or different spin multiplicities, (i.e. singlet–singlet, or triplet–triplet), as long as, during the transfer, the excited and ground states do not change multiplicity (figure 7.11). The singlet–singlet energy transfer may involve the Coulomb interaction but the triplet–triplet transfer does not, since that would violate the spin conservation rule.

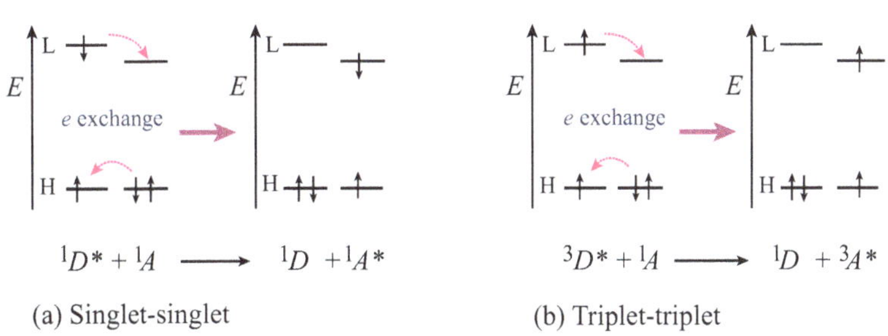

Figure 7.11. Energy level diagram for the Dexter energy transfer processes in the contact zone. (a) Singlet–singlet. (b) Triplet–triplet.

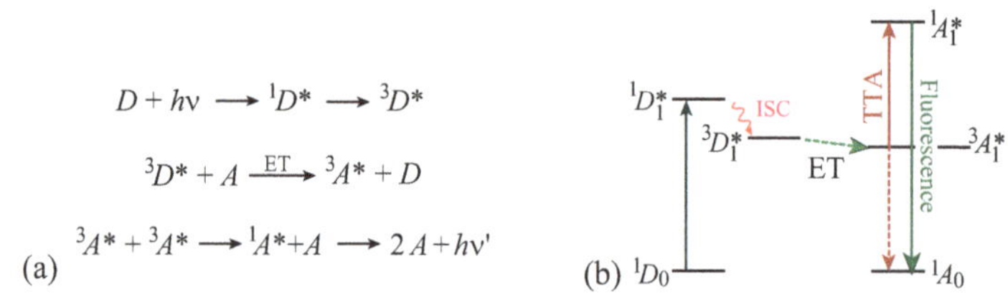

Figure 7.12. (a) Triplet–triplet annihilation process (TTA). (b) Simplified energy level diagram, and states involved in a TTA process enabled by a donor–acceptor pair (D–A).

The rate of energy transfer decays exponentially with the donor–acceptor distance, R_{DA}:

$$k_{ET,Exc} \propto (1/R_{DA}^2)\exp(-2R_{DA}/L)\, J,$$

where L is the sum of the van der Waals radii of D and A, and J is the spectral overlap.

Another important process, particularly for applications in solar cells, is triplet–triplet annihilation (TTA), in which two triplets annihilate each other and generate a singlet. The process is schematically depicted in figure 7.12.

In this process D is excited into a singlet excited state, ${}^1D_1^*$, followed by ISC into a triplet excited state ${}^3D_1^*$. A Dexter-type triplet–triplet energy transfer (ET) then occurs between the ${}^3D_1^*$ state and a triplet excited state of the acceptor, ${}^3A_1^*$. When two molecules in ${}^3A_1^*$ states collide, TTA may occur, with one molecule relaxing to the ground state, ${}^1A_0$, and the other being upconverted to a higher singlet excited state, ${}^1A_1^*$, eventually followed by fluorescence to the ground state.

Identification of chemical names mentioned in the text

ITIC	3,9-bis(2-methylene-(3-(1,1-dicyanomethylene)-indanone))-5,5,11,11-tetrakis (4-hexylphenyl)-dithieno[2,3-*d*:2′, 3′-*d*′]-*s*-indaceno[1,2-*b*:5,6-*b*′]dithiophene
PPV	poly(p-phenylene vinylene)
CN-PPV	cyano-polyphenylene vinylene
MEH-PPV	poly(2-methoxy-5-(2′-ethyl-hexyloxy)-1,4-phenylene vinylene)
PEDOT:PSS	poly(3,4-ethylenedioxythiophene) poly(styrenesulfonate)
P3HT	poly(3-hexylthiophene-2,5-diyl)
HT-P3HT	head to tail (HT) poly(3-hexylthiophene-2,5-diyl)

Further reading

Alcácer L 2018 *Electronic Structure of Organic Semiconductors* (San Rafael, CA: Morgan & Claypool)

Forrest S R 2020 *Organic Electronics: Foundations to Applications* (Oxford: Oxford University Press)

Hadziioannou G and Malliaras G G (ed) 2007 *Semiconducting Polymers: Chemistry, Physics and Engineering* 2nd edn (New York: Wiley) (two-volume set)
Kelley A M 2013 *Condensed-phase Molecular Spectroscopy and Photophysics* (Hoboken: Wiley)
Köhler A and Bässler H 2015 *Electronic Processes in Organic Semiconductors: An Introduction* (Weinheim: Wiley)
Lanzani G 2012 *The Photophysics behind Photovoltaics and Photonics* (Weinheim: Wiley)
Reynolds J R, Thompson B C and Skotheim T A (ed) 2019 *Conjugated Polymers. Properties, Processing, and Applications* 4th edn (Boca Raton, FL: CRC Press)

Problems

7.1 Consider the lattice structure of the HT-P3HT unit cell shown in figure 7.4.
   (a) Part (b) shows four units like the one represented in (a) associated with the unit cell, but there are only two chains per unit cell. Is there any inconsistency?
   (b) In the context of the figure, explain what constitutes a polymer chain and why the two chains in the unit cell are non-equivalent.
   (c) Calculate the volume of the unit cell.
   (d) Determine the density of HT-P3HT in this phase.

7.2 A particular III–V semiconductor has a dielectric constant of 10 and effective electron and hole masses of 0.05 m_e and 0.5 m_e, respectively.
   (a) Calculate the energies of the first three Wannier exciton levels.
   (b) Calculate the radius of the 1*s* exciton.

7.3 Consider a donor–acceptor pair D–A, whose dipole moments are parallel, in a medium with dielectric constant $\varepsilon_r = 2.6$. Assume that the absorption coefficient of the acceptor is constant over the spectral overlap region, with a value of $\varepsilon_A = 10^5 \mathrm{cm^{-1} mol^{-1} L}$ at a wavelength of 600 nm, and that the fluorescence quantum yield is 0.95.
   (a) Calculate the Förster radius for this D–A system.
   (b) Assuming that the efficiency of energy transfer is 50%, calculate the D–A distance.

References

[1] Tang C W 1986 *Appl. Phys. Lett.* **48** 183–5
[2] Tsumura A, Koezuka H and Ando T 1986 *Appl. Phys. Lett.* **49** 1210–2
[3] Tang C W and Van Slyke S A 1987 *Appl. Phys. Lett.* **51** 913–5
[4] Burroughes J H, Bradley D D C, Brown A R, Marks R N, Mackay K, Friend R H, Burns P L and Holmes A B 1990 *Nature* **347** 539–41
[5] Braun D and Heeger A J 1991 *Appl. Phys. Lett.* **58** 1982–4
[6] Ohmori Y, Uchida M, Muro K and Yoshino K 1991 *Jpn. J. Appl. Phys.* **30** L1941–43
[7] Braun D, Gustafsson G, McBranch D and Heeger A J 1992 *J. Appl. Phys.* **72** 564–8
[8] Yu G and Heeger A J 1995 *J. Appl. Phys.* **78** 4510–5

[9] Halls J J M, Walsh C A, Greenham N C, Marseglia E A, Friend R H, Moratti S C and Holmes A B 1995 *Nature* **376** 498–500
[10] Baldo M A, O'Brien D, You Y, Shoustikov A, Sibley S, Thompson M and Forrest S 1998 *Nature* **395** 151–4
[11] Akamatu H and Inokuchi H 1950 *J. Chem. Phys.* **18** 810–1
[12] Kearns D and Calvin M 1958 *J. Chem. Phys.* **29** 950–1
[13] Shirakawa H, Louis E J, MacDiarmid A G, Chiang C K and Heeger A J 1977 *J. Chem. Soc., Chem. Commun.* **16** 578–80
[14] Bolto B, McNeill R and Weiss D 1963 *Aust. J. Chem.* **16** 1090–103
[15] Kanazawa K K, Diaz A F, Geiss R H, Gill W D, Kwak J F, Logan J A, Rabolt J F and Street G B 1979 *J. Chem. Soc., Chem. Commun.* 854–5
[16] Ivory D M, Miller G G, Sowa J M, Shacklette L W, Chance R R and Baughman R H 1979 *J. Chem. Phys.* **71** 1506–7
[17] Hernandez R, Diaz A F, Waltman R and Bargon J 1984 *J. Phys. Chem.* **88** 3333–7
[18] Chiang J C and MacDiarmid A G 1986 *Synth. Met.* **13** 193–205
[19] Bryce M R, Chissel A, Kathirgamanathan P, Parker D and Smith N R M 1987 *J. Chem. Soc., Chem. Commun.* **6** 466–7
[20] Zhu Q, Fischer J E, Zusok R and Roth S 1992 *Solid State Commun.* **83** 179–83
[21] Hudson B S 2018 *Materials* **11** 242
[22] Bredas J L 2014 *Mater. Horiz.* **1** 17–9
[23] Sworakowski J 2018 *Synth. Met.* **235** 125–30
[24] Campbell R B, Robertson J M and Trotter J 1961 *Acta Cryst.* **14** 705–11
[25] Campbell R B, Robertson J M and Trotter J 1962 *Acta Cryst.* **15** 289–90
[26] Holmes D, Kumaraswamy S, Matzger A J and Vollhardt P C 1999 *Chem. Eur. J* **5** 3399–412
[27] Mattheus C C, Dros A B, Baas J, Meetsma A, Boer J L d and Palstra T T M 2001 *Acta Crystallogr.* C **57** 939–41
[28] Mattheus C C, Dros A B, Baas J, Oostergetel G T, Meetsma A, de Boer J L and Palstra T T M 2003 *Synth. Met.* **138** 475–81
[29] Fritz S E, Martin S M, Frisbie C D, Ward M D and Toney M F 2004 *J. Am. Chem. Soc.* **126** 4084–5
[30] Ruiz R, Mayer A C, Malliaras G G, Nickel B, Scoles G, Kazimirov A, Kim H, Headrick R L and Islam Z 2004 *App. Phys. Lett.* **85** 4926–8
[31] Schiefer S, Huth M, Dobrinevski A and Nickel B 2007 *J. Am. Chem. Soc.* **129** 10316–7
[32] Yoshida H, Inaba K and Sato N 2007 *Appl. Phys. Lett.* **90** 181930
[33] Natsume Y, Minakata T and Aoyagi T 2009 *Synth. Met.* **159** 338–42
[34] Mannsfeld S C B, Virkar A, Reese C, Toney M F and Bao Z 2009 *Adv. Mater.* **21** 2294–8
[35] Yoneya M, Kawasakib M and Ando M 2010 *J. Mater. Chem.* **20** 10397–402
[36] Cornil J, Calbert J P and Brédas J L 2001 *J. Am. Chem. Soc.* **123** 1250–1
[37] Troisi A and Orlandi G 2005 *J. Phys. Chem.* B **109** 1849–56
[38] Parisse P, Ottaviano L, Delley B and Picozzi S 2007 *J. Phys.: Condens. Matter.* **19** 106209
[39] Zhu Z, Wu S and Zhang Y 2008 *Russ. J. Phys. Chem.* A **82** 2293–8
[40] Endres R G, Fong C Y, Yang L H, Witte G and Wöll C 2004 *Comput. Mater. Sci.* **29** 362–70
[41] Chiang C K, Fincher C R, Park Y W, Heeger A J, Shirakawa H, Louis E J, Gau S C and MacDiarmid A G 1977 *Phys. Rev. Lett.* **39** 1098–101
[42] Swager T M 2017 *Macromolecules* **50** 4867–86
[43] MacDiarmid A G 2001 *Angew. Chem. Int. Ed.* **40** 2581–90

[44] Vernitskaya T V and Efimov O N 1997 *Russ. Chem. Rev.* **66** 443–57
[45] Lee K, Cho S, Heum Park S, Heeger A J, Lee C W and Lee S H 2006 *Nature* **441** 65
[46] McCullough R D, Tristram-Nagle S, Williams S P, Lowe R D and Jayaraman M 1993 *J. Am. Chem. Soc.* **115** 4910–1
[47] Kayunkid N, Uttiya S and Brinkmann M 2010 *Macromolecules* **43** 4961–7
[48] Colle R, Grosso G, Ronzani A and Zicovich-Wilson C M 2011 *Phys. Status Solidi* B **248** 1360–8
[49] Dudenko D, Kiersnowski A, Shu J, Pisula W, Sebastiani D, Spiess H W and Hansen M R 2012 *Angew. Chem.* **124** 11230–4
[50] Brinkmann M, Hartmann L, Kayunkid N and Djurado D 2014 Understanding the structure and crystallization of regioregular poly (3-hexylthiophene) from the perspective of epitaxy *P3HT Revisited—From Molecular Scale to Solar Cell Devices* (Advances in Polymer Science vol 265) ed S Ludwigs (Berlin, Heidelberg: Springer) ch 3 pp 83–106
[51] Böckmann M, Schemme T, de Jong D H, Denz C, Heuer A and Doltsinis N L 2015 *Phys. Chem. Chem. Phys.* **17** 28616–25
[52] Tremel K and Ludwigs S 2014 Morphology of P3HT in Thin Films in Relation to Optical and Electrical Properties *P3HT Revisited – From Molecular Scale to Solar Cell Devices* (Advances in Polymer Science vol 265) ed S Ludwigs (Berlin, Heidelberg: Springer) ch 2 pp 39–82
[53] Bin H, Zhang Z G, Gao L, Chen S, Zhong L, Xue L, Yang C and Li Y 2016 *J. Am. Chem. Soc.* **138** 4657–64
[54] Greenham N C, Moratti S C, Bradley D D C, Friend R H and Holmes A B 1993 *Nature* **365** 628–30
[55] Halls J J M, Cornil J, dos Santos D A, Silbey R, Hwang D H, Holmes A B, Brédas J L and Friend R H 1999 *Phys. Rev.* B **60** 5721–7
[56] Yu P, Zhen Y, Dong H and Hu W 2019 *Chem* **5** 2814–53
[57] Su W P, Schrieffer J R and Heeger A J 1979 *Phys. Rev. Lett.* **42** 1698
[58] Su W P, Schrieffer J R and Heeger A J 1980 *Phys. Rev.* B **22** 2099–111
[59] Bredas J L and Street G B 1985 *Acc. Chem. Res.* **18** 309–15
[60] Heeger A J, Kivelson S, Schrieffer J R and Su W P 1988 *Rev. Mod. Phys.* **60** 781–850
[61] Fisher A J, Hayes W and Wallace D S 1989 *J. Phys. Condens. Matter* **1** 5567
[62] Holstein T 1959 *Ann. Phys. (N. Y.)* **8** 343–89
[63] Holstein T 2000 *Ann. Phys. (N. Y.)* **281** 725–73
[64] Coropceanu V, Cornil J, da Silva Filho D A, Olivier Y, Silbey R and Brédas J L 2007 *Chem. Rev.* **107** 926–52
[65] Mott N and Davis E A 1979 *Electronic Processes in Non-Crystalline Materials* Monographs on Physics 2 (Oxford: Oxford University Press)
[66] Nardes A M, Kemerink M, Janssen R A J, Bastiaansen J A M, Kiggen N M M, Langeveld B M W, van Breemen A J J M and de Kok M M 2007 *Adv. Mater.* **19** 1196–200
[67] Smith N V 2001 *Phys. Rev.* B **64** 155106
[68] Ter Haar D 1967 *The Old Quantum Theory* (The Commonwealth and International Library: Selected Readings in Physics vol 1) (Oxford: Pergamon) pp 167–83 Translation of 'On the Quantum Theory of Radiation', 1917, by A. Einstein
[69] Jablonski A 1933 *Nature* **131** 839
[70] Permogorov S 1975 *Phys. Status Solidi* B **68** 9–42
[71] Zhu X Y, Yang Q and Muntwiler M 2009 *Acc. Chem. Res.* **42** 1779–87

[72] Friend R H, Bradley D D C and Townsend P D 1987 *J. Phys. D: Appl. Phys.* **20** 1367
[73] Campbell I H, Hagler T W, Smith D L and Ferraris J P 1996 *Phys. Rev. Lett.* **76** 1900–3
[74] Conwell E M 1996 *Synth. Met.* **83** 101–2
[75] Brédas J L, Cornil J and Heeger A J 1996 *Adv. Mater.* **8** 447–52
[76] Liu X, Li Y, Ding K and Forrest S 2019 *Phys. Rev. Appl.* **11** 024060
[77] Förster T 1948 *Ann. Phys.* **437** 55–75
[78] Dexter D L 1953 *J. Chem. Phys.* **21** 836–50
[79] Baldo M A, Thompson M E and Forrest S R 2000 *Nature* **403** 750–3
[80] Baldo M A and Forrest S R 2000 *Phys. Rev.* B **62** 10958–66

Chapter 8

Organic electronic devices

This chapter deals with the applications of organic semiconductors, which are already numerous and still increasing as new properties are discovered and devices are invented. A detailed description of the most common devices is presented. Particular attention is given to organic light-emitting devices, organic solar cells, organic field-effect transistors, organic electrochemical transistors, and special applications in bioelectronics.

8.1 Introduction

Organic semiconductors form the basis of an overwhelming variety of applications, ranging from light-emitting devices and solar cells to most types of electronic components and sensors. The most remarkable features of these materials are their lightness and possible transparency, which makes them suitable for large-area and flexible electronics. Some of these applications allow for solution-processing methods, with the possibility of large-scale roll-to-roll printing on flexible substrates. Applications in the life sciences, such as body-integrated sensors, wearable biomedical systems, healthcare monitors, and implantable electronics, are already part of the so-called 'organic bioelectronics' field, and it is expected that, in a foreseeable future, they will support high levels of autonomy, taking advantage of artificial intelligence-based advanced medical diagnostics [1].

Organic light-emitting diodes (OLEDs) were the first and so far the most successful organic electronic devices to reach the market on a large scale. They are presently used in active matrix displays for television screens, computer monitors, and smartphones and are gaining prominence in the lighting market. On the other hand, organic solar cells (OSCs) are a promising low-cost new generation of photovoltaic energy conversion devices.

The main drawback of organic semiconductors is their low charge mobility when compared with those of inorganics, which limits the range of applications to low-processing-speed devices. On the other hand, organic semiconductors are unique in

doi:10.1088/978-0-7503-3347-4ch8

many aspects, such as diversity, softness, low-cost processability and, in many cases, transparency and biocompatibility. Purity may be a problem, but it is not as important as in inorganic semiconductors for some applications, such as solar cells, because the relevant processes occur at molecule-to-molecule contacts (e.g. donor–acceptor contacts in bulk heterojunctions).

8.2 Organic light-emitting diodes

The first OLED structure, introduced by Tang and Van Slyke in 1987 [2], consisted of a double layer of organic thin films: a hole transport layer (an aromatic diamine) of about 75 nm, and an electroluminescent layer of the fluorescent metal complex Alq_3 of about 60 nm. The anode was an indium tin oxide (ITO) thin film on a glass substrate, and the cathode was a Mg:Ag alloy deposited by co-evaporation. The OLED could be operated in a continuous dc or pulsed mode and behaved as a rectifier, with a forward voltage of a few volts. The light output was proportional to the current (in the range of 0.1 to 100 mA cm^{-2}). The external quantum efficiency (photons/electron) was about 1%, the luminous efficacy was 1.5 lm W^{-1}, and the brightness was greater than 1000 cd m^{-2} at driving voltages of less than 10 V. The electroluminescence (EL) spectrum, with a peak intensity at 550 nm (green), was identical to the photoluminescence spectrum of Alq_3. The first polymer OLED (or PLED) using PPV, which was prepared using a solution-processable precursor, was reported in 1990 by Richard Friend and co-workers, opening a promising future for large-area OLED manufacture [3].

The architecture of a typical OLED and the corresponding energy level diagram and electroluminescence generation processes are represented in figure 8.1. Electroluminescence in organic semiconductors is the result of the radiative recombination of excitons created by the combination of electrons (negative polarons) and holes (positive polarons), which are generated by electron and hole injection from a low-work-function metal, acting as the cathode, and a high-work-function transparent conductor, acting as the anode, respectively. It essentially has four basic processes: (1) charge injection, (2) charge transport via polaron hopping between molecular sites or from chain to chain (in polymers), (3) exciton

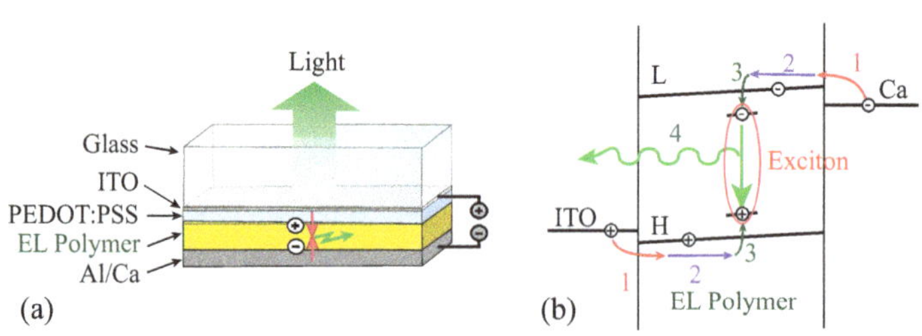

Figure 8.1. (a) Architecture of an OLED. An electroluminescent material is deposited on a glass substrate covered with ITO and a hole injection layer (PEDOT:PSS). The cathode is a Ca thin film protected by an Al layer. (b) Energy level diagram and electroluminescence generation processes in an OLED: (1) Charge injection (electrons and holes), (2) charge transport, (3) exciton creation, (4) exciton recombination and emission. H and L denote the HOMO and LUMO of the electroluminescent organic material.

creation (formation of bound *e–h* pairs), and (4) exciton recombination and light emission.

The choice of materials and architecture is crucial to each type of application and processing technology. In today's highly competitive display market, electroluminescent small molecules are usually preferred, along with vapour deposition technologies. Conjugated polymers can be solution processed, but this limits device complexity due to the fact that the solvents used may harm the underlying layers. Conjugated polymers may be better fitted for large-area, low-cost, high-flexibility applications. The full range of colours can be achieved by an appropriate choice of electroluminescent materials. Naturally, the electronic parameters, such as the work functions of the cathode and anode, the energies of the HOMO and LUMO levels of the photoactive materials, and any additional transport and blocking layers, if needed, have to match in such a way that they can perform their functions. Figure 8.2 shows the main types of layers in a typical multilayer OLED structure; note that in many instances, one layer plays the role of two or three layers, and that there may be additional dipole layers at interfaces to adjust energy barriers.

The hole transport layer (HTL) has to match the hole injection layer (HIL), which should provide efficient hole injection from the anode. The electron blocking layer (EBL) should have a wide energy gap and a LUMO level higher than the LUMO level of the electron emission layer (EML) in order to block electrons leaving the emission layer. Conversely, the electron injection layer (EIL) should provide efficient electron injection from the cathode, which should have a work function matching the LUMO of the EIL. A hole blocking layer with a wide gap and a low HOMO may be needed to prevent holes from escaping the EML. The EML is the electroluminescent layer, which can be a small molecule-type semiconductor, a conjugated polymer, a polymer blend, etc. In many cases the EML is a host–guest mixture of the electroluminescent material and a charge transport material. In such cases, energy transfer (Förster, Dexter, or radiative) usually occurs from the host to the guest molecule.

Among the conceivable architectures, apart from the most common, which are based on a transparent anode (e.g. ITO) on a transparent substrate, we can consider

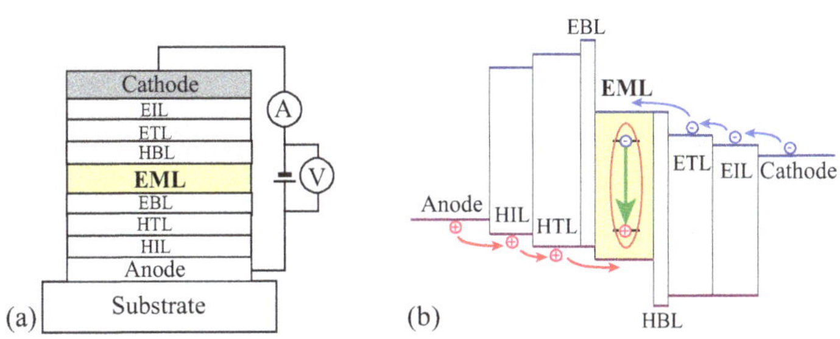

Figure 8.2. (a) Multilayer structure of a typical OLED and respective energy level diagram (b). From anode to cathode the following layers are shown: hole injection layer (HIL), hole transport layer (HTL), electron blocking layer (EBL), electron emission layer (EML), hole blocking layer (HBL), electron transport layer (ETL), and electron injection layer (EIL). The upper horizontal lines indicate the cathode work function and the LUMO levels of the materials. The lower lines indicate the HOMO levels and the anode work function.

OLEDs that emit through transparent metallic cathodes (with thicknesses of less than the skin depth) in which non-transparent substrates are used, as well as inverted OLEDs in which the cathodes are deposited on the supporting substrates. Such architectures may be beneficial for some lighting and display applications.

8.2.1 Electroluminescent processes

Currently, three main types of emissive materials are available that support radiative exciton recombination: fluorescent, phosphorescent, and thermally activated delayed fluorescence (TADF) materials. A schematic overview of these mechanisms is shown in figure 8.3.

Each electron–hole pair generates a singlet exciton ($S = 0$; $M_S = 0$) with a 25% probability or a triple exciton ($S = 1$; $M_s = +1$, $M_s = 0$, $M_s = -1$) with a 75% probability (see also figure 7.8). The spin selection rule $\Delta S = 0$ for electronic transitions implies that only $S_1 \rightarrow S_0$ is allowed, which corresponds to fluorescence (figure 8.3(a)). This selection rule may be relaxed in systems with a high degree of spin–orbit coupling (e.g. from heavy atoms), allowing intersystem crossing (ISC) leading to phosphorescence (figure 8.3(b)). When the energy difference ΔE_{ST} between the T_1 and S_1 states is small, triplets may be endothermically promoted into the singlet manifold, followed by thermally activated delayed fluorescence emission, TADF, as shown in figure 8.3(c).

Emission from the singlet state, S_1, can occur in conventional fluorescence (figure 8.3(a)), leading to devices with internal quantum efficiencies (IQEs, photons generated per electron) that cannot exceed 25%, as in the case of the first OLED reported by Tang *et al* [2]. The fluorescence quantum yield, η_F, which is related to the IQE, is given by

$$\eta_F = k_F/(k_F + k_{NRR} + k_{ISC}), \qquad \tau_F = 1/(k_F + k_{NRR} + k_{ISC}),$$

where k_F is the radiative rate constant, k_{NRR} is the non-radiative rate constant, and k_{ISC} is the ISC rate constant. In OLEDs, τ_F is in the nanosecond range. k_F can be determined by measurements of the fluorescence quantum yield, η_F, and the fluorescence lifetime, $\tau_F = \eta_F/k_F$. Typical examples of fluorescent OLED materials are Alq_3, rubrene, and DPVBi.

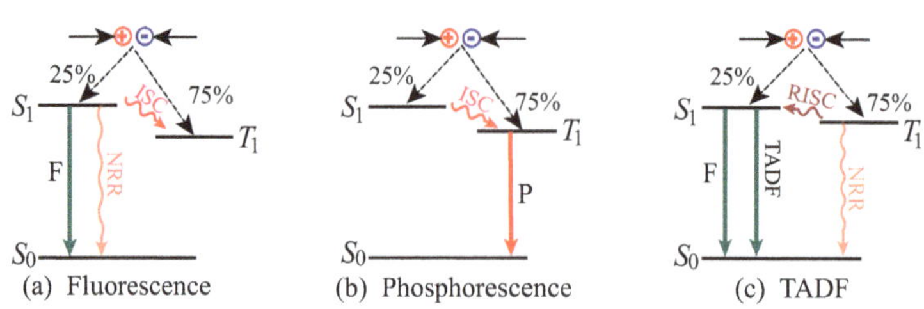

Figure 8.3. Schematic showing the mechanisms of fluorescence (F), phosphorescence (P) and thermally activated delayed fluorescence (TADF). The energy difference between the S_1 and T_1 states is of the order of 0.5–1 eV for the fluorescent (a) and phosphorescent cases (b), while it is usually less than 0.1 eV for the TADF case (c). ISC, RISC, and NRR stand for intersystem crossing, reverse intersystem crossing, and non-radiative relaxation, respectively. The 25% and 75% probabilities are derived from spin statistics.

In addition to fluorescence, singlet emission can also occur as a result of at least three other processes: thermally activated delayed fluorescence (TADF), triplet–triplet annihilation (TTA), and hybridized local and charge transfer (HLCT). In the TADF process (also called E-type delayed fluorescence) the first excited singlet state S_1 becomes populated by a thermally activated reverse intersystem crossing (RISC) mechanism from the first excited triplet T_1. In this process all triplet excitons may lead to delayed fluorescence, in addition to the singlet excitons, which produce prompt normal fluorescence. For the upconversion of the RISC process to be active at room temperature, the activation energy ΔE_{ST} must be smaller than 0.2 eV. An example of high-efficiency fluorescent TADF OLEDs using a phosphorescent sensitizer, enabled by Förster energy transfer (FRET), has been reported by Baldo *et al* [4]. TTA (see figure 7.11) consists of the fusion of two triplet excitons into one singlet excited state. A TTA emitter contains a prompt component (conventional fluorescence) and a delayed fluorescence component that results from the TTA mechanism. A maximum IQE value of 62.5% can, in principle, be attained [5]. Hybridized local and charge-transfer excited state emission is another approach to harvesting the 75% non-emissive triplet excitons [6] through RISC from upper triplet levels (hot excitons) to the singlet manifold in appropriate donor–acceptor material systems, simultaneously harvesting singlet and triplet excitons.

Emission originating from spin doublets, rather than singlet or triplet excitons, has been reported for radical-based OLEDs [7, 8]. These radical emitters (e.g. TTM-1Cz) have a singly occupied molecular orbital (SOMO) in the ground state, corresponding to an $S = 1/2$ doublet. When the electron in the SOMO is excited to the lowest singly unoccupied molecular orbital (SUMO), the SOMO becomes empty. The transition of the excited electron back to the SOMO is 100% spin-allowed. Electroluminescence in OLEDs originating from the electron transition between the SUMO and the SOMO has been reported with a maximum external quantum efficiency (EQE, photons emitted per electron) of 27% at a wavelength of 710 nm, which is the highest reported value for deep-red and infrared LEDs [8].

Emission from triplet states may occur as phosphorescence, overcoming the forbidden transition from T_1 to S_0, by mixing the S_1 and T_1 states through ISC enabled by strong spin–orbit coupling due to the presence of a heavy atom. This process, which can be achieved by doping a host molecule with an organometallic complex, uses both singlet and triplet excited states (figure 8.3(b)), opening the way for internal quantum efficiencies near 100%. The first examples of efficient phosphorescent OLEDs were reported by Baldo *et al* [9, 10]. One of their devices had the following structure: ITO/NPD/Ir(ppy)$_3$:CBP/BCP/Alq$_3$/MgAg, where NPD is the hole transport layer, Ir(ppy)$_3$:CBP is the emission layer formed by the green electrophosphorescent iridium complex, Ir(ppy)$_3$, doped into CBP, BCP is a hole blocking barrier (essential to confine excitons within the EML and achieve high IQE), and Alq$_3$ acts as an electron transport layer. ITO and MgAg are the transparent anode and cathode, respectively. A luminance of 100 cd m^{-2} and external quantum and power efficiencies of 7.5% (26 cd A^{-1}) and 19 lm W^{-1}, respectively, were obtained at an applied voltage of 4.3 V. Such high performance is explained by efficient transfer of both singlet and triplet excitons from the host to Ir(ppy)$_3$.

There is evidence that the whole emission comes from the triplet state, supported by the fact that the transient response of $Ir(ppy)_3$ in CBP has a mono-exponential phosphorescence decay of ~500 ns with an absence of fluorescence from the $Ir(ppy)_3$, indicating strong spin–orbit coupling and ISC from S_1 to T_1. In addition, emission from CBP is negligible even at high current densities, indicating complete energy transfer [10].

Phosphorescent polymer OLEDs have been extensively reported, specifically using phosphorescent dyes doped into charge-transporting polymer matrices (see Further reading).

8.2.2 Characterization

OLEDs are primarily characterized by their EQE, η_{ext}, and *power efficiency* (PE), η_P. The EQE is the ratio of the number of photons emitted by the OLED to the number of electrons injected. The EQE may be decomposed into a product of the efficiencies of several elementary processes:

$$\eta_{ext} = \eta_{int}\, \eta_{out}, \qquad \eta_{int} = \gamma\, \eta_{PL}\, \chi_{ST}\,.$$

η_{int} is the IQE, i.e. the ratio of the number of photons generated to the number of injected electrons. This factor includes all photons generated in the emission layer (note that some will be lost by absorption or reflection from the interfaces). η_{out} is the outcoupling (or light output) efficiency, i.e. the fraction of photons that finally leaves the device.

The contributions to η_{int} include:

1. The recombination efficiency, γ, a loss factor that accounts for the imperfect balance of electrons and holes in the emission layer, recombination via charge traps, etc.
2. The photoluminescence quantum yield, η_{PL}, of the emission layer. Note that the radiative recombination probabilities are assumed to be identical for singlet excitons produced both optically and electrically .
3. The fraction of spin-allowed excitons, χ_{ST}, i.e. the ratio of excitons generated for each *e–h* pair injected into the emission layer. From the considerations presented in the previous section, we can conclude that if γ and η_{PL} are 100%, the maximum IQEs, η_{int}, of fluorescent, phosphorescent, and TADF OLEDs can reach 25%, 100%, and 100%, respectively.

The outcoupling efficiency, η_{out}, depends on the refractive index n of the substrate, and is of the order of $1/2n^2 \approx 0.2$, for glass ($n \approx 1.5$). Based on this argument, the maximum EQEs of fluorescent, phosphorescent, and TADF OLEDs are about 5%, 20%, and 20%, respectively, if no additional light extraction improvement technique is used. Various methods of enhancing η_{out} have been proposed, including the implementation of special structures to avoid reflection at interfaces, and micro- and nanostructures to release trapped photons.

The power efficiency, which measures the ratio of the power *in* to the power *out*, is given by

$$\eta_P(J) = \eta_{ext}\frac{P_{EL}}{J\ V} = \eta_{ext}\frac{\langle h\nu\rangle}{e\ V(J)}$$

where P_{EL} is the total emitted electroluminescent power density, J is the OLED injection current density at voltage V, $\langle h\nu\rangle$ is the average energy of the emitted photons, e is the elementary charge, and $V(J)$ is the applied voltage at current density J.

Radiometric and photometric quantities

In addition to the quantum and power efficiencies, several other quantities are used to quantify the performance of OLEDs. These quantities are either *radiometric* or *photometric*. Radiometric quantities are electromagnetic performance instrumental measures (e.g. using photodetectors and spectrometers), while photometric quantities are perceptual characteristics, such as the colour and brightness of the source as perceived by the (normal) human eye.

The main radiometric quantities are: (i) *power* (or radiant flux), Φ_e, measured in watts (W), (ii) *irradiance*, E_e, measured in W m^{-2}, (iii) *radiant intensity*, I_e, measured in watts per steradian (the steradian, or square radian, sr, is the SI unit of solid angle), W sr^{-1}, and (iv) *radiance*, L_e, measured in W m^{-2} sr^{-1}. The photometric units are based on the SI unit for *luminous intensity*, the candela, cd. Other units based on the candela are, for example, the *luminance*, L, expressed in cd m^{-2}, the *luminous flux*, Φ_V, measured in lumens, lm (1 lm = 1 cd sr), the *illuminance*, E, whose SI unit is the lux, lx (1 lx = 1 lm m^{-2}), and the *efficacy* (also known as the *luminous power efficiency*) measured in lm W^{-1}.

The radiometric and photometric quantities are related by the photopic spectral luminous efficiency function $V(\lambda)$, which is the spectral response of the human eye to the various wavelengths of light. $V(\lambda)$ is a standard curve tabulated by the 'Commission Internationale de l' Éclairage' (CIE). The curve has a maximum value of unity at the wavelength of 555 nm (green), which is the wavelength at which the human eye is most sensitive. By definition, the candela, cd, is the luminous intensity, in a given direction, of a source that emits monochromatic radiation with a frequency of 540×10^{12} hertz ($\lambda = 555$ nm) and that has a radiant intensity in that direction of 1/683 watt per steradian. In other words, in a candela, there are 683 lumens per watt for 555 nm light. To convert from watts to lumens, for a monochromatic light source, simply multiply the power in watts by the appropriate $V(\lambda)$ value and the conversion factor (683 lm W^{-1}) from the definition of the candela.

The following equation calculates the total luminous flux in a source of light

$$\Phi_V = K_m \int_\lambda \Phi_E(\lambda)\ V(\lambda)\ d\lambda$$

where Φ_V is the luminous flux in lumens, K_m is the scaling factor, 683 lm W^{-1}, $\Phi_E(\lambda)$ is the spectral power in W nm^{-1}, and $V(\lambda)$ is the photopic spectral luminous efficiency curve. Since $V(\lambda)$ is given by a table of empirical values, the integration is usually numerical. Table 8.1 summarizes the relation between the radiometric and photometric power units. State-of-the-art OLEDs can produce light with nearly

Table 8.1. Radiometric and photometric units.

Quantity	Radiometric unit	Photometric unit
Power	W	lm, cd sr
Power per unit area	W m^{-2}	lm m^{-2}, cd sr m^{-2}, lx
Power per unit solid angle	W sr^{-1}	cd
Power per unit area per solid angle	W m^{-2} sr^{-1}	cd m^{-2}, lm m^{-2} sr^{-1}

every colour in the CIE chromaticity diagram, but standard procedures for the manufacture of full-colour displays or lighting devices have not yet been established.

8.2.3 White OLEDs for general lighting

The colour quality of an illumination source is characterized by the colour rendering index (CRI) and the correlated colour temperature (CCT). The CRI measures how closely a light source matches the black body radiation at a particular temperature T. The highest possible CRI value is 100, which corresponds to a source identical to standardized daylight. This is based on the assumption that the Sun is an ideal black body with a surface temperature of 5780 K. Real white corresponds to $T = 6000$ K, while the so-called 'cool white' corresponds to $T > 5000$ K, and 'warm white' to T between 2700 K and 3000 K. The CCT is the corresponding black body temperature (e.g. $CCT = 6500$–5000 K for standard daylight light sources).

The first white OLED (WOLED) was reported by Kido *et al* in 1995 [11]. It was made from vacuum-deposited organic thin films with three emitting layers, each emitting red, green, or blue light, to generate white light. A luminance of more than 2000 cd m^{-2} (almost that of a fluorescent lamp) has been reported at voltages of the order of 15–16 volts.

Various techniques are currently under development, such as blending the red, green, and blue emitting molecules, so that each emits in the appropriate proportion to provide illumination with the desired colour qualities. The highest OLED efficacies (luminous power efficiencies, $\eta_{LP} = E/JV$) are in the range of 120–150 lm W^{-1}, and have CRIs of the order of 95 (lab demonstrators). This should be compared with the incandescent bulb (17 lm W^{-1}), the fluorescent lamp (80–90 lm W^{-1}), and the inorganic LED (80–90 lm W^{-1}, 240 lm W^{-1} in a lab demo).

OLED lighting is a promising emergent technology with many options for materials, both small molecules and polymers. Costs are still too high, but they will certainly come down as mass production and standardization become significant.

8.3 Organic solar cells

The photovoltaic effect was discovered in 1839 by the 19-year-old Alexandre-Edmond Becquerel, the father of Antoine Henri Becquerel who shared the Nobel Prize in Physics with Pierre and Marie Curie for the discovery of radioactivity. However, the first practical solar cell was only developed in 1954 by Chapin, Fuller, and Pearson at Bell Labs; it was a silicon cell with 6% efficiency [12]. The first organic solar cell with a significantly high fill factor, of a new type based on a molecular junction, was disclosed by Ching Tang at Kodak in 1986 [13]. OSCs are the topic of this section. Here, we will try to give a concise overview of the main aspects of the physics of this promising type of solar cell.

8.3.1 Basic operating principles

OSCs, like OLEDs, are diodes; they are based on similar processes, but operating in reverse order. While OLEDs emit light by the recombination of excitons formed from injected charge carriers, solar cells generate charge carriers by the dissociation of excitons created by light absorption. Figure 8.4 illustrates the photogeneration processes in a donor–acceptor bilayer solar cell and in a bulk heterojunction (BHJ) solar cell under illuminated and short-circuit conditions. There are four main steps: (1) the absorption of photons and the creation of excitons (for convenience, we will use the terms electron, hole, and exciton for the respective quasiparticles, including polaronic effects); (2) exciton diffusion towards the heterojunction; (3) transfer of the electron to the acceptor, while the hole remains in the donor, forming the charge-transfer (CT) state (if the exciton is formed in the acceptor, there is a hole transfer to the donor); (4) dissociation of the *e–h* pair, whose former members drift away from the junction towards the electrodes. In the BHJ cell, the donor is blended with the acceptor throughout the whole film. Thus, photogenerated excitons can dissociate into charges at any point within the blend.

The EQE, $EQE(\lambda)$ or $\eta_{\text{ext}}(\lambda)$, is defined as the ratio of the number of collected electrons, N_e, to the number of incident photons, $N_{\text{ph}}(\lambda)$, of a given wavelength λ. The ideal $\eta_{\text{ext}}(\lambda)$ function has a square shape, with constant values across the whole range of wavelengths:

$$\eta_{\text{ext}}(\lambda) = N_e/N_{\text{ph}}(\lambda). \tag{8.1}$$

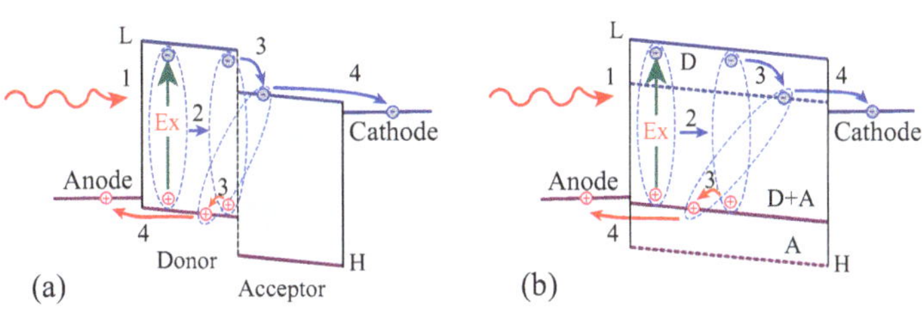

Figure 8.4. Energy level diagrams of (a) a bilayer solar cell and (b) a BHJ solar cell. H and L denote the HOMO and LUMO levels of the donor (D) and the acceptor (A), respectively. The generation of charge carriers can be summarized in four steps: (1) the absorption of photons and exciton creation; (2) exciton diffusion; (3) electron transfer from the donor to the acceptor at the heterojunction (HJ) and formation of the CT state; (4) dissociation of the *e–h* pair, whose former members drift away from the junction towards the electrodes. The use of 'anode' and 'cathode' to designate the electrode with the 'higher work function' and the 'lower work function', respectively, is derived from OLEDs. In organic solar cells, the proper designation is 'cathode' and 'anode', respectively, as used for a battery.

The IQE, $IQE(\lambda)$ or $\eta_{int}(\lambda)$, is the efficiency with which photons that are absorbed into the active layer (AL) can generate electrons that can be collected:

$$\eta_{int}(\lambda) = \frac{N_e}{N_{ph/AL}(\lambda)} = \frac{N_e}{N_{ph}(\lambda)} \times \frac{N_{ph}(\lambda)}{N_{ph/AL}(\lambda)} = \frac{\eta_{ext}(\lambda)}{Abs_{AL}(\lambda)},$$

where Abs_{AL} is the absorption in the active layer.

The IQE is the product of the efficiencies of each of the four steps shown in figure 8.4:

$$\eta_{int} = \eta_A \eta_D \eta_T \eta_C,$$

where η_A is the efficiency of absorption and exciton creation (step 1), η_D is the exciton diffusion efficiency (step 2), η_T is the efficiency of electron transfer from the donor to the acceptor (step 3), and η_C is the efficiency of the *e–h* pair dissociation and the drift of charge carriers towards the electrodes for charge collection (step 4). The efficiency of each of these processes is less than 100%, and therefore the overall internal quantum efficiency may be very small. However, it can be much improved by a wise choice of materials and architectures, especially by controlling the morphology at the HJ, and reach nearly 100% as, for example, in [14].

The performance of a solar cell is measured by the *power conversion efficiency* (PCE or η_P), which is defined as the ratio of the maximum electrical power density output, P_M (in W m^{-2}), to the incident power density from the Sun, P_{in},

$$\eta_P = P_M / P_{in}.$$

The standard illumination is usually taken as 1000 W m^{-2}, or 1 'Sun', and corresponds to the standard spectrum at the Earth's surface, AM1.5G, defined by the American Society for Testing and Materials (ASTM) for flat plate modules (AM and G stand for air mass and global, respectively) and includes direct and diffuse light.

An OSC can be modelled as a current source in parallel with a diode, as shown in figure 8.5(a); a shunt resistance R_{SH} (slope of $J–V$ at $V = 0$) and a series resistance, R_S, which is the sum of the contact and internal resistances, are included. The power conversion efficiency, η_P, is calculated directly from the current–voltage characteristics of the solar cell in the dark and under illumination represented in figure 8.5(b), where V_{OC} is the open-circuit voltage and J_{SC} is the short-circuit current density.

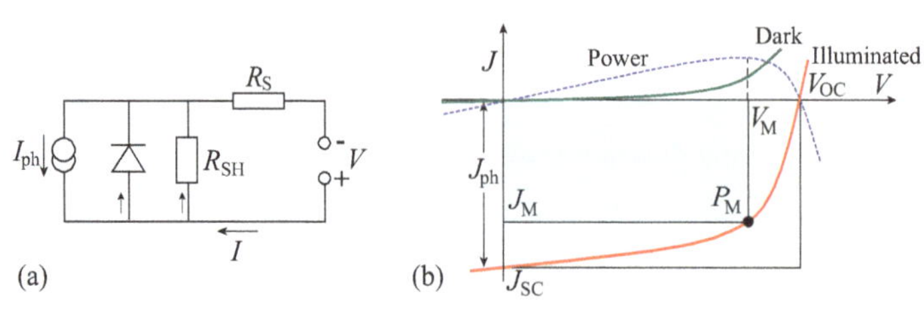

Figure 8.5. (a) Equivalent circuit for an OSC. The diode represents the solar cell in the dark and I_{ph} denotes the photogenerated current. (b) Current–voltage ($J–V$) characteristic in the dark and under illumination. J is the current density, i.e. the current I divided by the area of the cell.

The maximum power density output, P_M, is determined by the point at which the product of the voltage and the current density is maximal. The ratio of P_M to the product of V_{OC} and J_{SC} is the *fill factor* (FF).

In the dark, the current is given by the diode equation $I_d = I_S[\exp(eV/n\, k_B T) - 1]$, where e is the elementary charge, and n is the ideality factor (the deviation from ideal diode characteristics) which varies between one and two. I_S is the 'dark saturation current' (leakage current in the dark). Under illumination, the current flows in reverse bias: electrons flow to the lower-work-function metal (usually called the cathode) and holes flow to the higher-work-function metal (usually called the anode). The power density output is maximal at $P_M = J_M\, V_M$. The power conversion efficiency can then be written as:

$$PCE \equiv \eta_P = \frac{P_M}{P_{in}} = FF\,\frac{J_{SC}\,V_{OC}}{P_{in}}, \qquad FF = \frac{J_M\,V_M}{J_{SC}\,V_{OC}}. \tag{8.2}$$

On the basis of the equivalent circuit (figure 8.5(a)), the I–V characteristics of the solar cell can then be written as:

$$\begin{aligned} I &= I_d + I_{SH} - I_{ph} \\ &= I_S\left[\exp\left(\frac{e\,V}{n\,k_B T}\right) - 1\right] + I_{SH} - I_{ph} \\ &= I_S\left\{\exp\left[\frac{e}{n\,k_B T}(V - IR_S)\right] - 1\right\} + \frac{V - I\,R_S}{R_{SH}} - I_{ph}, \end{aligned}$$

where I_{SH} is the current through the shunt resistance R_{SH}. R_{SH} should be very large to prevent leakage currents (short circuits, pinholes, traps, etc). I_{ph} is the photocurrent.

The short-circuit current density, J_{SC}, can be derived from the EQE, $\eta_{ext}(\lambda)$, as follows:

$$J_{SC} = e\int_0^{\infty} \Phi_{Sun}(\lambda)\,\eta_{ext}(\lambda)\,d\lambda,$$

where e is the elementary charge and $\Phi_{Sun}(\lambda)$ is the photon flux (photons $m^{-2}\,s^{-1}$) corresponding to the AM1.5G solar spectrum. The EQE (equation 8.1) can be estimated by measuring the short-circuit current of a calibrated photodiode with a known EQE and then measuring the short-circuit current of the device of unknown *EQE*, since $\eta_{ext,unk}(\lambda) = [I_{unk}(\lambda)/I_d(\lambda)]\eta_{ext,d}(\lambda)$, where the subscripts 'unk' and 'd' stand for the unknown device and the calibrated photodiode, respectively.

The theoretical limits of solar cell efficiency have been studied since the mid 1950s, but the widely accepted reference is the 1961 paper by Shockley and Queisser [15], derived for inorganic solar cells, based on the assumption that the solar cell behaves as a black body, and that all interactions of photons with the solar cell occur through the generation and recombination of free *e–h* pairs. In addition, it is presumed that all photons with energies above the gap are absorbed, and that the charge carriers have infinite mobility. The total number of generated electrons, N_e, can be calculated from the $\eta_{ext}(\lambda)$ curve, assuming that all photons, each with energy $E_{ph} = hc/\lambda$, can be converted into electrons. The maximum voltage (which depends on the gap) is given by $V_{max} = E_{ph}(\lambda_{max})/e$, and the Shockley–Queisser limit can be written as

$$\eta_{P_{max}} = \frac{V_{max}(\lambda)\, J(V_{max})}{P_{in}} = t_s\, u(x_g)\, v(fx_g\, x_c)\, FF(fx_g\, x_c),$$

where t_s is the probability that photons with energy above the gap, $h\nu_g$, incident from the Sun's black body spectrum create an e–h pair; $u(x_g) = h\nu_g\Phi_{abs}/P_{in}$ (where Φ_{abs} is the absorbed photon flux density) is the power conversion efficiency given by the ratio of the output power density to the incident power density; $v(fx_g\, x_c) = eV_{oc}/h\nu_g$ relates the open-circuit voltage to the optical gap; f is a combination of factors that includes a geometrical factor (see [15] for details); x_c is the ratio of the cell's temperature to the Sun's temperature, T_c/T_s; and $x_g = h\nu_g/k_BT$. Detailed balance limit tables relating the semiconductor gap to the solar cell's characteristics and software to perform calculations can easily be found in the literature.

The maximum PCE that can be reached for a Si p–n junction solar cell is ca. 33%, assuming an AM1.5 solar spectrum and a temperature of 300 K. For organic solar cells the PCE depends on both the optical gap and the exciton binding energy (in the range 0.3–1 eV), which is neglected in inorganic solar cells (with exciton binding energies of the order of 5–10 meV). These and other factors imply that the power conversion efficiencies of organic solar cells are intrinsically lower than those predicted by the Shockley–Queisser theory. A thermodynamic efficiency limit for excitonic solar cells was derived by Giebink *et al* [16], who generalized the Shockley–Queisser model to include sizeable exciton binding energies, which are more appropriate for organic solar cells; their model recovers the Shockley–Queisser results in the limit of zero exciton binding energy. The predicted values fall in the range of 22%–27%.

The equivalent circuit and the equations for the inorganic solar cells can be used, but V_{OC}, which is the maximum voltage that a solar cell can deliver to the external circuit, is one of the most important parameters that determine the PCE, as shown in equation (8.2). V_{OC} is determined by the work functions of the electrode materials and/or by the HOMO and LUMO levels of the donor and acceptor, respectively. An in-depth review discussing the many factors that influence the V_{OC} and strategies for improving it was produced by Elumalai and Uddin [17]. J_{SC} depends on the absorption efficiency, charge generation, D–A charge transfer, charge carrier mobility, morphology, and the area of the D/A interface at the HJ. The FF depends on many factors, including the contacts and the morphology.

The high exciton binding energy in organic semiconductors poses a problem for dissociation. This is overcome, at least in part, by the use of the D–A heterojunction (identified as a type-II heterojunction) strategy. At the D/A interface, the Coulomb potential well of the electron (due to the e–h interaction) extends into the acceptor, which has a larger electron affinity than the donor, and the electron may tunnel into the conduction band of the acceptor. If the electron affinity of the electron in the acceptor is large enough, the electron will have sufficient kinetic energy to overcome the Coulomb attraction and separate from the hole, which stays in the donor.

Exciton dissociation is a fundamental question and is still a matter of controversy. A widely accepted explanation for the higher-than-expected IQE of some types of organic solar cells is based on the so-called 'hot exciton' effect. The question is: does dissociation occur before or after exciton thermalization? Grancini *et al* [18] examined the photophysical processes at the D/A interface of an efficient low-bandgap polymer/$PC_{60}BM$ solar cell and found that high-energy (hot) singlet excitons generate interfacial CT states and free polarons within the first tens of femtoseconds after photoexcitation (much faster than thermalization processes) and therefore hot exciton dissociation occurs before thermalization. In this way, the photon energy above the optical gap is not lost, which leads to an increase in the IQE. This hypothesis has been challenged by Noriega *et al* [19] who claimed that no extra excitation energy is needed to increase the rate of exciton dissociation, and that the improvement of the IQE is due to the existence of networks of disordered semicrystalline aggregates of particular sizes that enable better charge transport along the crystalline domains. The charge-transfer yield certainly depends on the structural order of the blend at the D/A interface, as demonstrated by Herrmann *et al* [20], but the role of hot excitons in the increase of the D–A charge-transfer yield has been confirmed by several experiments and theoretical calculations, such as those reported in [21–24]. Most probably, a combination of both structural and hot exciton effects contributes to the increase of the IQE.

8.3.2 Architectures and materials

The architectures and materials of organic solar cells have to take into account the high values of the exciton binding energy and the low mobility of the charge carriers in organic semiconductors. As discussed above, the first problem is solved by using donor–acceptor (D–A) heterojunctions, and the second by an appropriate choice of materials. Possible architectures for OSCs can be inferred from the energy level diagrams of figure 8.4. In bilayer OSCs, the donor and acceptor layers are sandwiched between two electrodes with different work functions. The cathode's work function should match the LUMO level of the acceptor and the anode's work function should match the HOMO level of the donor in order to achieve good electron and hole collection. The heterojunction can be planar or structured in a way that maximizes the interface area and the IQE. The morphology and thickness of the layers and the structure of the interface should be 'designed' by taking into account not only the matching of the relevant energy levels of the intervening materials but also the diffusion length of the excitons (a critical parameter of the order of 10–20 nm) and the charge mobilities of the donor and acceptor. The transparent anode is generally ITO, and the cathode is generally a low-work-function metal, such as Al, Mg, or Ca. There are also additional layers, such as a hole extraction layer (usually PEDOT:PSS), an eventual electron transport layer, or a thin layer of LiF (<1 nm) to improve the contact efficiency (increasing the V_{OC}). A recent review of interfacial materials, including hole and electron transport layers for organic solar cells, is provided in reference [25]. In bilayer heterojunctions, the donor and acceptor layers are in contact with the anode and cathode, respectively. The D/A interface can be nanostructured with an interpenetrating network, for example, using nanoimprint lithography [26] or another method as described in reference [27], to obtain interdigitated segregated domains (e.g. columns) of the donor and acceptor materials that are perpendicular to the anode and cathode surfaces, respectively, with domain widths that are within the range of the exciton diffusion lengths (ideally ca. <10 nm in polymers). An advantage of the bilayer heterojunction, if properly built, is that electrons and holes travel separately in the acceptor and in the donor, respectively, reducing charge recombination.

In BHJ OSCs, the donor and acceptor are blended, but are segregated in different phases and form an interpenetrating D/A interface network dispersed over the whole bulk; the resulting large interfacial area is where charge transfer and exciton dissociation occur. The morphology and dimensions of the segregated D and A phases should be such as to enable exciton diffusion into the D/A interface within the exciton lifetime, and it is important that the charges separated within their proper phases should find percolated pathways to the respective electrodes.

Besides the conventional architecture, OSCs can be inverted (with the charges flowing in opposite directions), or built as three-layer p–i–n-like structures with an interlayer between the hole transport layer and the electron transport layer. OSCs can also be combined in tandem by stacking two or more heterojunction solar cells in order to use a broader range of the solar spectrum.

The choice of materials for OSCs depends on the intended application (e.g. indoor, outdoor, large-area, mobile systems, medical), the type of technology (e.g. solution processing, inkjet or roll-to-roll printing, vacuum deposition), considering that OSCs can be flexible, lightweight, transparent, placed on anything, anywhere, and use direct or diffuse light.

As a first approach, the way to choose the donor–acceptor pair is to determine whether the fluorescence efficiency of one of the components is strongly reduced when the second is added, or as a result of photoinduced absorption (absorption of the excited states).

Both donor and acceptor materials can be small molecules or polymers. Electron donors with wide gaps, such as MEH-PPV ($\Delta E_o = 2.30$ eV), have reduced light absorption and low mobility. Donor and acceptor materials with low bandgaps lead to greater light absorption, which increases exciton yield and charge carrier generation. D–A copolymers (based on D and A moieties) have been used to build the backbone of the donor material, and more recently D–A–D polymers (i.e. polymers with a D–A–D 'mer', or unit) have successfully been used with matching acceptors to obtain highly efficient OSCs, as for example in [28].

Wide-gap polymer donors have recently been designed to match some small-gap acceptors, such as ITIC (figure 7.5) and Y6 (see figure 8.6) [29]. An 18% efficient polymer solar cell has been reported with the structure ITO/PEDOT:PSS/D18:Y6/PDIN/Ag [28], in which the D18 D–A–D polymer ($\Delta E_o = 2.74$ eV) was combined with Y6 ($\Delta E_o = 1.55$ eV). As explained by Liu *et al* [30], the high performance of the small-gap non-fullerene acceptors (1.3–1.6 eV) with broad absorption, such as Y6, is

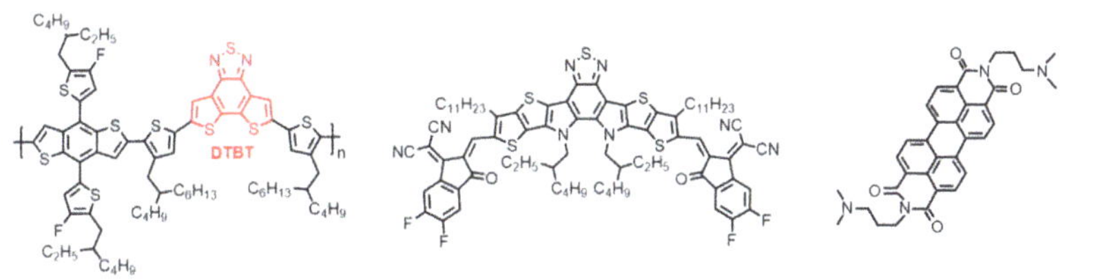

D18: E_H=-5.51; E_L=-2.77 Y6: E_H=-5.65; E_L=-4.10 PDIN: E_H=-6.05; E_L=-3.72

Figure 8.6. Chemical structure of D18, based on a donor–acceptor–donor 'mer', the Y6 acceptor, and PDIN, an electron-transporting layer material, used in an 18% efficient polymer solar cell, with the structure ITO/PEDOT:PSS/D18:Y6/PDIN/Ag [28].

due to high conjugation length, good compatibility between donor and acceptor, adjusted crystallinity, and device engineering.

The acceptor choice (small molecule or polymer) is as important as the choice of the donor material, since both have to be combined to make a heterojunction. The present trend is focused on non-fullerene materials which, for example, can be processed using so-called green solvents [31]. Thick-film OSCs are of major importance for many applications and processing technologies. The topic has recently been reviewed by Chang *et al* [32].

8.4 Organic transistors

Transistors based on organic semiconductors can be of different types. The most common fall under the generic name of 'organic field-effect transistors' (OFETs). OFETs generally adopt the architecture of thin-film transistors based on low-mobility inorganic semiconductors, such as amorphous silicon, and they also work in *accumulation mode*.

An OFET can be viewed as a plane capacitor in which one of the conductive plates is the *gate* and the other is the organic semiconductor, which is in direct contact with two electrodes, the *source* and the *drain*. A dielectric layer separates the gate from the semiconductor and the two electrodes, as shown in figure 8.7(a).

When a voltage V_G is applied between the source (assumed to be connected to ground) and the gate, charges (electrons or holes) are induced at the semiconductor/dielectric interface, forming (above a given threshold voltage) a thin conducting *channel* with a conductance proportional to V_G. This gate-induced charging is called the *field effect*. At low drain voltages, V_D, the current I_D increases linearly with V_D. When the drain voltage reaches a certain value compared to the gate voltage, the channel is pinched off, and the current enters the saturation regime, becoming independent of the source–drain voltage. A plot of the I_D current vs the gate voltage, V_G, at a constant drain voltage is called the *transfer curve*. This plot gives the threshold voltage, V_T, and the mobility of the charges in the channel.

In the following discussion, we assume that the channel is p-type (i.e. the majority carriers in the semiconductor are holes). For an n-type channel, the polarities would simply be reversed. The surface charge density (charge per unit area in $C\,cm^{-2}$) induced at the position x in the channel by the applied gate voltage, V_G, is given by

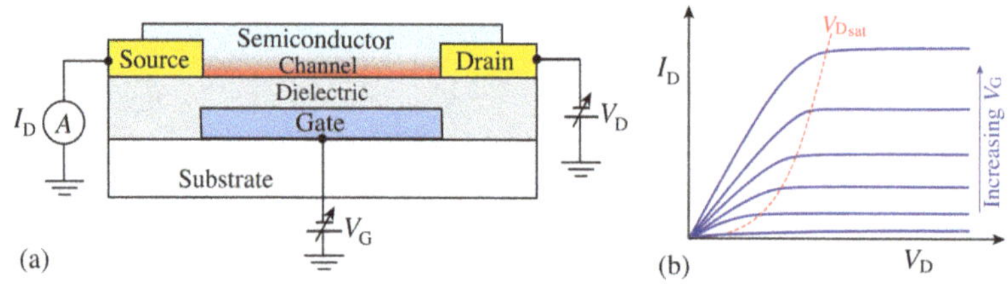

Figure 8.7. (a) Common organic field-effect transistor geometry (bottom gate, bottom contacts). (b) Current–voltage (I_D–V_D) characteristics for various gate voltages, V_G. The dotted line represents the voltage at the onset of saturation.

$$Q_{ind}(x) = n(x)e\ t = C_i\left[V_G - V_T - V(x)\right], \tag{8.3}$$

where $n(x)$ is the charge number density (in cm^{-3}); e is the elementary charge; t is the thickness of the charged layer; C_i is the capacitance of the dielectric per unit area, ($C_i = \varepsilon_0\varepsilon_r/t$); V_T is the *threshold voltage* (the voltage at which charges effectively start to be induced); and $V(x)$ is the voltage at point x. When the drain voltage V_D is non-zero but less than V_G, there is a linear gradient in the charge density along the channel, therefore, for small values of V_D, it is a reasonable approximation to take the average value, which corresponds to the voltage at the midpoint in the channel. This average value is then

$$Q_{ind} = C_i(V_G - V_T - V_D/2),$$

from which we can derive the I–V characteristics for the transistor. By Ohm's law, the drain current is given by $I_D = (\sigma A/L)V_D$, where σ is the conductivity; A is the section, i.e. the product of the width and the thickness of the channel, $A = Wt$; and L is the channel length. Considering that $\sigma = n\mu e$, where μ is the mobility, we can write $\sigma = Q_{ind}\,\mu/t$. We can then write the drain current as

$$I_D = (W/L)\mu\ C_i[(V_G - V_T) - V_D/2]V_D. \tag{8.4}$$

This is the equation commonly used to describe the so-called *linear regime*, in which $V_D < (V_G - V_T)$. At $V_D = V_G - V_T$, the channel becomes pinched, meaning that the region of the channel near the drain is depleted of charge carriers, and the I–V characteristics reach the *saturation regime*. If we substitute $V_D = V_G - V_T$ into (8.4), we get

$$I_{D_{sat}} = \mu\frac{WC_i}{2L}(V_G - V_T)^2, \tag{8.5}$$

from which we can extract the mobility by plotting $\sqrt{I_{D_{sat}}}$ vs V_G at a given V_D and calculate the slope of the linear region. In addition, we can get the threshold voltage, V_T, by extrapolation of the linear region to zero current. The mobility calculated in the saturation regime is indeed a mean value, since the charge density is nonuniform along the channel and the mobility depends on the charge density. As an alternative, we can calculate the mobility in the linear regime of the output I_D–V_D characteristics, in which the charge density is more uniform, using the *transconductance*, g_m (in units of $[A\ V^{-1}] = [S]$), defined as

$$g_m = \left.\frac{\partial I_D}{\partial V_G}\right|_{V_D} = \frac{W}{L} C_i \, \mu \, V_D, \qquad V_D \leqslant V_{D_{sat}}.$$

This simple method has some drawbacks, because the mobility may depend on the gate voltage and other factors. More accurate methods are described in the recommended literature.

In addition to the mobility and the threshold voltage, two other parameters are often used to evaluate the transistor's performance: one is the ON/OFF current ratio, I_{ON}/I_{OFF}, which can be taken from the transfer curve I_D vs V_G in the saturation regime at high V_G. I_{ON} can depend on numerous factors, such as the mobility of the semiconductor, the capacitance of the gate dielectric, the operating voltages, and the density and type of trap states present at the semiconductor/dielectric interface. The OFF current is the leakage current when V_G is zero, and may be due to eventual current paths through the bulk. ON/OFF ratios as high as 10^8 and field-effect mobilities in the range of 0.02–0.05 $cm^2\ V^{-1}\ s^{-1}$ have been reported for OFETs based on bisdithienothiophene (BDT), a molecule with a particularly suitable crystal structure and thin film morphology formation [33]. In fact, in BDT, the planes of adjacent molecules along the π–π stacking are exactly parallel (not herringbone-type), providing a strong π–π face-to-face overlap. In addition, the long axes of the molecules in the domains of the polycrystalline film are perpendicular to the substrate.

In any one of the four architectures (bottom gate, bottom contacts; bottom gate, top contacts; top gate, bottom contacts; and top gate, top contacts) there are two important interfaces: the one between the gate insulator (dielectric) and the semiconductor, and that between the electrodes (source and drain) and the semiconductor. Although it is often used as as an insulator, SiO_2 is not adequate for OFETs, because it tends to lead to the formation of silanol (Si–O–H) groups which act as traps for electrons; this is one reason that organic semiconductors are usually *p*-type [34]. This can be avoided by introducing a self-assembled monolayer (SAM) with adequate dipole moments between the dielectric and the semiconductor, but the tendency is to use dielectric polymers. The interface between the semiconductor and the electrodes determines the contact resistance, which depends on the height of the energy barrier between the electrodes and the semiconductor and is the difference between the metal work function and the ionization potential of the semiconductor for the injection of holes (or the electron affinity for the injection of electrons). Also, in this case, the use of SAMs may improve the contact resistance. Actually, it is now clear that it is not only the nature of the semiconductor (p- or n-type) that determines the nature of the channel; for this reason, it is more appropriate to refer to p- or n-type channel transistors [35].

8.4.1 Organic light-emitting transistors

Organic light-emitting transistors (OLETs) combine, in the same architecture, the radiative recombination of electrons and holes of electroluminescent organic semiconductors (such as OLEDs) with the field effect of OFETs. They are therefore bifunctional devices with light emission and switching capabilities. The first OLET was reported by Hepp *et al* [36] and was essentially a thin-film polycrystalline tetracene OFET with the possibility of injecting both holes and electrons, which recombined at some point within the channel, leading to light emission. Although unipolar organic transistors can behave as light emitters under some specific conditions, it is reasonable to expect that efficient OLETs should be ambipolar, and possibly multilayer, with, for example, a hole transport layer, an electron transport layer, and an emissive layer in between, in an appropriate architecture that promotes the formation of excitons and light emission. In an ideal ambipolar OLET, the channel conducts electrons and holes for opposite polarities of the gate voltage. When the applied gate voltage lies between the source and the drain voltages, electron and hole accumulation layers form in the channel near the respective electrodes, as depicted in figure 8.8. The electrons and holes are expected to recombine and emit light where the two layers meet.

In a seminal and comprehensive article, Zaumseil *et al* [37] demonstrated spatially controlled light emission from an ambipolar OLET based on a solution-processed conjugated polymer (OC_1C_{10}-PPV), in an architecture such that electrons were injected from a calcium electrode and holes from a gold electrode; they recombined radiatively at a position in the channel which could be controlled (and visualized) by the applied gate and source–drain voltages. The mobilities were extracted from the transfer curves in the saturated regime (equation 8.5) for different negative and positive source–drain voltages, V_D, and found to be 3×10^{-3} cm^{2}V^{-1}s^{-1} for electrons, and 6×10^{-4} cm^{2}V^{-1}s^{-1} for holes, respectively. To derive an expression for the position of the light emission in the channel, the authors assumed that the total length of the channel was the sum of the hole and electron channel lengths, $L_{\text{total}} = L_h + L_e$, and used equation (8.5), considering that $I_h = I_e = I_D$ and that in the electron channel, the effective gate voltage was $V_G - V_D$. They obtained the following equation for the ratio of the hole and electron channel lengths:

$$\frac{L_h}{L_e} = \frac{\mu_h}{\mu_e} \frac{(V_G - V_{T_h})^2}{(V_G - V_D - V_{T_e})^2}$$

where the indices 'h' and 'e' refer to holes and electrons, respectively. Since the EQE, η_{ext}, is the ratio between the number of photons emitted and the number of electrons

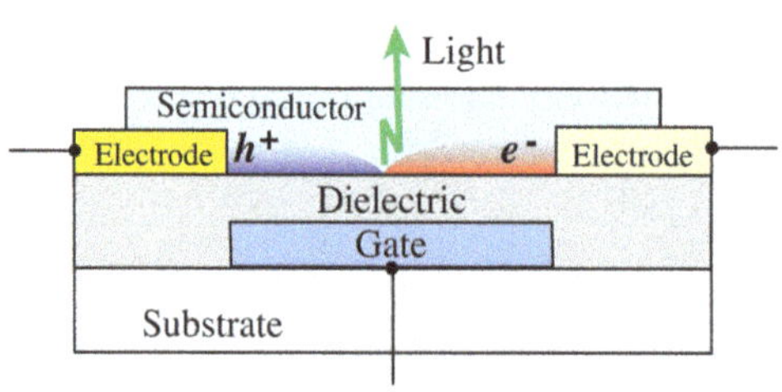

Figure 8.8. Schematic representation of an ambipolar organic light-emitting transistor. Light emission occurs where the electron and hole accumulation layers meet. In ambipolar transistors the nomenclature 'source' and 'drain' is ambiguous, since electrons and holes may be injected from either electrode, which is why the term 'electrode' is used in this figure.

injected, $\eta_{ext} = (P_{tot}/E_{ph})/(I_D/q)$, where P_{tot} is the total light power emitted and E_{ph} is the average photon energy (q is the electron charge), we can write η_{ext} as follows, if we express the power in [W] and the energy in [eV] [6]:

$$\eta_{ext} = \frac{P_{tot}}{I_D\, E_{ph}} = \frac{I_{ph}/K}{I_D\, E_{ph}} \approx 1.5\,\frac{I_{ph}}{I_D}$$

where K is the photo sensitivity of the detector in A W^{-1} (e.g. 0.3 A W^{-1}), I_{ph} is the photocurrent, and I_D is the drain current of the transistor.

The interest in OLETs has significantly increased in recent times due to their potential applications as light sources that combine electroluminescence with switching in the same device. OLETs also have a potential use in the development of electrically pumped lasers. The most promising OLETs are multilayer devices with complex architectures. Capelli *et al* [38] reported a p-channel/emitter/n-channel trilayer heterostructure, with a higher efficiency (EQE>1%) than those of the equivalent OLEDs. The specific details of the materials were as follows: the gate, the dielectric, and the source and drain were ITO, PMMA, and Au, respectively, while the active region was a superposition of three layers: the first, in contact with the dielectric, was an electron transport layer (DFH-4T), the light-emitting layer was Alq_3:DCM, and DH-4T was a hole transport layer, which was in contact with the source and the drain. To give a further example, Chen *et al* [39] reported multilayer rigid and flexible OLETs based on high-dielectric-constant crosslinked and modified poly(vinyl alcohol) dielectrics and a guest–host system with a phosphorescent emitter, which had EQEs in the range of 7.3%–9% and brightnesses in the range of 8300–14500 cd m^{-2}.

In spite of the progress in the development of efficient devices, an OLET-based technology is unlikely to emerge in the near future, due to the lack of suitable materials and architectures. Two recent review articles are worth reading, namely that of Chaudhry *et al* [40] and that of Qin *et al* [41].

8.4.2 Organic electrochemical transistors

Organic electrochemical transistors (OECTs) are analogous to organic field-effect transistors in which the dielectric is replaced by an electrolyte. In addition, the active semiconductor must conduct both electronic (electrons and/or holes) and ionic charges. PEDOT:PSS, which is a hole conductor, is the most popular semiconductor choice. The first OECT was reported by White *et al*, who used polypyrrole as the active semiconductor to fabricate a microelectrode array that works as a transistor when immersed in an electrolyte [42]. Figure 8.9(a) represents the general concept of an OECT, in which the active semiconductor layer forming the channel is deposited on a substrate and is in contact with the source and drain. The channel is also in contact with an electrolyte in which the gate is immersed. Figure 8.9(b) shows the schematic layout of a lateral OECT architecture similar to that described by David Nilsson [43], but using inkjet-printed PEDOT:PSS patterns to work as the source, channel, and drain of a connected structure, the gate being separated from the

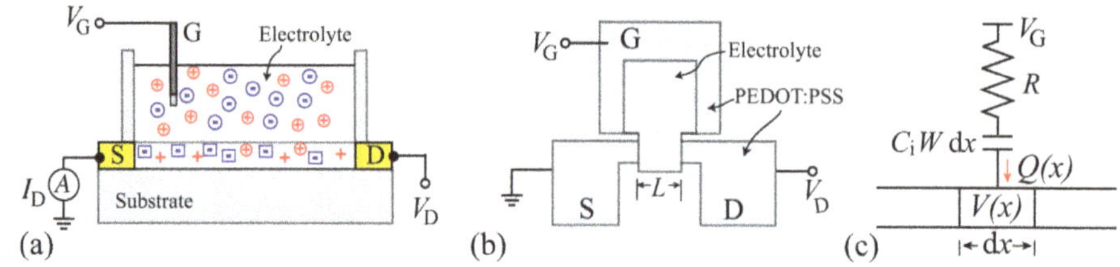

Figure 8.9. (a) Schematic representation of an organic electrochemical transistor (OECT). (b) Layout of a PEDOT:PSS printed lateral OECT. Adapted from [43], copyright Wiley-VCH Verlag GmbH, reproduced with permission. (c) Coupling of the ionic circuit to the electronic circuit at a position x along the channel. Adapted from [45], copyright 2007 WILEY-VCH Verlag GmbH & Co, used with permission.

channel by a small gap. A layer of a highly conductive polymer electrolyte covers part of the gate and the channel, defining its length, L, [44].

Figure 8.9(c) represents the coupling of the ionic circuit to the electronic circuit, indicating the charge $Q(x)$ that is injected from the electrolyte into the channel at position x.

OECTs do not operate in the same way as OFETs. Not only is the capacitance describing the coupling of the gate to the channel dependent on volume (i.e. dependent on the unit volume, rather than the unit area), but also, when a positive voltage is applied to the gate, the ionic charges of the electrolyte are believed to enter and diffuse within the whole volume of the channel, compensating for the charges of the sulphonate anions ($-SO_3^-$) of the PSS (in the case of a PEDOT:PSS OECT).

The most popular ways to describe the operation of OECTs are based on the Bernards and Malliaras model [45], which was proposed to address the working principles of PEDOT:PSS OECTs in depletion mode by considering the devices formed by an electronic circuit and used an ionic circuit based on the ideal polarizable electrode theory to explain ion transport in the electrolyte. The steady-state and transient characteristic parameters can be extracted from this model for both positive and negative drain voltages when a positive gate voltage is applied ($V_G > 0$). The electronic circuit was described in a way that differs from that used for OFETs, as it considers the channel capacitance to depend on volume rather than per unit area as in equation (8.3), due to the fact that the electronic charges (holes) present in the whole volume of the channel are partly complemented by the ionic charges of the cations that are injected from the electrolyte into the channel. Instead of going through the derivation of the original model or its variants, we prefer to follow the simple and comprehensive approach to the Bernards and Malliaras model used by Rivnay *et al* [46] (Supplementary Materials) for PEDOT:PSS OECTs, which starts by applying Ohm's law to an element dx of the channel length:

$$I_D(x) = \sigma \, dV(x)/dx = e\mu \, p(x) \, dV(x)/dx, \qquad (8.6)$$

where $\sigma = e\mu p(x)$ is the conductivity; e is the elementary charge; μ is the hole mobility; and W and t are the width and thickness of the channel, respectively; $dV(x)/dx$ is the electric field; and $p(x)$ and $V(x)$ are the hole density and voltage at position x in the channel. The hole density, $p(x)$, is equal to the number density of sulphonate anions ($-SO_3^-$) (compensated for by p_o holes) in the pristine PEDOT:PSS, minus the number density, $M^+(x)$, of cations that enter the channel when a positive voltage V_G is applied to the gate, i.e.

$$p(x) = p_o - M^+(x) = p_o - (C_v/e)\left[V_G - V_T - V(x)\right], \tag{8.7}$$

where we expressed $M^+(x)$ in terms of the volume-dependent channel capacitance, C_V. By substituting (8.7) into (8.6) and integrating over the channel length, L, we obtain the drain current,

$$I_D = \frac{Wt}{L}\mu C_v(V_T - V_G + V_D/2)V_D,$$

where $V_T = ep_o/C_V$. When the channel is fully depleted at the drain, the drain current saturates. The saturation voltage is then $V_{D_{sat}} = V_G - V_T$, and the saturation current is

$$I_{D_{sat}} = -\frac{Wt}{2L}\mu C_v(V_T - V_G)^2 \tag{8.8}$$

from which we can extract the mobility by plotting $\sqrt{I_{D_{sat}}}$ vs V_G at a given V_D.

The transconductance at saturation is the derivative of (8.8) with respect to V_G:

$$g_{m_{sat}} = \frac{Wt}{L}C_V\,\mu\,(V_T - V_G),$$

from which we can also extract the hole mobility in the channel and other useful information.

While the electronic circuit of an OECT is modelled by the application of Ohm's law to the channel, the ionic circuit comprises a resistor R, which describes the ionic conductivity of the electrolyte, and a capacitor C in series, which describes the polarization at the gate/electrolyte and the electrolyte/channel interfaces.

When a constant voltage is applied to the gate, the current transient of the ionic circuit is described by the expression for the charging capacitor in the RC equivalent circuit,

$$Q(t) = Q_{ss}[1 - \exp(-t/\tau_i)],$$

where Q_{ss} is the total charge passing through the circuit in the steady state, and τ_i is the time constant $\tau_i = RC$. For more detailed information and an expansion of the Bernards and Malliaras model, including the performance of biosensors based on OECTs, we recommend the recent article by Nissa *et al* [48].

Organic electrochemical transistors have been of increasing interest in recent times for various reasons, namely because they can be manufactured in many different architectures, shapes, and sizes and for a large range of applications, particularly as transducers and amplifiers of biochemical signals that can work in aqueous environments and use aqueous, gel, or polymer electrolytes. PEDOT:PSS is by far the most popular organic semiconductor; it is ideal for the OECT channel, because it is a mixed electronic and ionic conductor, it can be made with a wide range of conductivities, it is transparent, and it is commercially available and biocompatible.

8.4.2.1 Organic bioelectronics and the applications of OECTs.
OECTs and OFETs have different properties and therefore different ranges of application. In terms of gain and frequency bandwidth, for example, OECTs show transconductances in the range of millisiemens, while OFETs usually operate in the range of microsiemens, but OFETs can operate at much higher frequencies (in the MHz band), whereas OECTs do not usually perform well above 1 kHz. Although OECTs have been applied in analogue and logic circuits in a variety of large-area and high-volume technologies, on a wide range of substrates, including paper and textiles, it will probably be in the broad field of organic bioelectronics that OECTs will have an advantage. Since they are soft materials that can transport both electronic and ionic charges, can work in aqueous solutions, and are biocompatible, they are already anticipating the dawn of a new era of devices and systems for applications in the life sciences and healthcare, in particular, for interfacing with the brain [49–53].

8.5 Conclusions

We have briefly described a few of the most common organic electronic devices, but there are many more that could be included, such as memories, electrical double-layer transistors, a great variety of sensors, drug delivery devices, ion pumps, and implantable electrodes and devices; and of course, electronic circuits and systems.

In organic electronic applications, there are many issues that need to be addressed. One of the utmost importance is the issue of the interfaces, particularly those involving dielectrics, electrolytes, and biological systems, a topic that is not addressed in this book. For this reason we highly recommend the review of ‘Interfaces in organic electronics’ by Fahlman *et al*, listed in Further reading.

Identification of chemical names mentioned in the text

Alq_3	tris(8-quinolinolato) aluminium
BCP	2,9-dimethyl-4,7-diphenyl-1,10-phenanthroline
CBP	4, 4′-bis(9-carbazolyl)-1, 1′-biphenyl
DCM	4-(dicyanomethylene)-2-methyl-6-(p-dimethylaminostyryl)-4H-pyran
DFH-4T	5, 5‴-bis(tridecafluorohexyl)-2, 2′: 5′, 2″: 5″, 2‴-quaterthiophene
DH-4T	3, 3‴-dihexyl-2, 2′: 5′, 2″: 5″, 2‴-quaterthiophene
DPVBi	4,4-bis(2,2-diphenylvinyl)-1,1-diphenyl
$Ir(ppy)_3$	tris[2-phenylpyridinato-C^2,N]iridium(III)
ITIC	3,9-bis(2-methylene-(3-(1,1-dicyanomethylene)-indanone))-5,5,11,11-tetrakis (4-hexylphenyl)-dithieno[2,3-*d*:2′, 3′-*d*′]-*s*-indaceno[1,2-*b*:5,6-*b*′]dithiophene.
MEH-PPV	poly(2-methoxy-5-(2′-ethyl-hexyloxy)-1,4-phenylene vinylene)
NPD	N, N′-di(1-naphthyl)-N, N′-diphenyl-(1, 1′-biphenyl)-4, 4′-diamine
OC_1C_{10}-PPV	poly(2-methoxy-5-(3,7-dimethyloctoxy)-p-phenylene-vinylene)
PCBM	[6,6]-phenyl-C_{61}-butyric acid methyl ester
PDIN	N, N′-bis[3-(dimethylamino)propyl]perylene-3,4,9,10-tetracarboxylic diimide
PEDOT: PSS	poly(3,4-ethylenedioxythiophene) poly(styrenesulfonate)
PMMA	poly(methyl methacrylate)

PPV	poly(p-phenylene vinylene)
TTM-1Cz	(4-N-carbazolyl-2,6-dichlorophenyl)bis(2,4,6-trichlorophenyl)-methyl

Further reading

Forrest S R 2020 *Organic Electronics: Foundations to Applications* (Oxford: Oxford University Press)

Halls J J M, Walsh C A, Greenham N C, Marseglia E A, Friend R H, Moratti S C, and Holmes A B 1995 *Nature* **376** 498–500 (https://doi.org/10.1038/376498a0). Important example to understand the processes in OLEDs and OSCs.

Hoppe H and Sariciftci N S 2004 Organic solar cells: an overview *J. Mater. Res.* **19** 1924–45

Kasap S and Capper P (ed) 2017 *Springer Handbook of Electronic and Photonic Materials* (Berlin: Springer)

Muccini M and Toffanin S 2016 *Organic Light-emitting Transistors: Towards the Next Generation Display Technology* (New Jersey: Wiley)

Tress W 2011 *Organic Solar Cells. Theory, Experiment, and Device Simulation* (Berlin: Springer)

Horowitz G 1998 Organic field-effect transistors *Adv. Mater.* **10** 365–77 (https://doi.org/10.1002/(SICI)1521-4095(199803)10:5<365::AID-ADMA365>3.0.CO;2-U)

Braga D and Horowitz G 2009 High-performance organic field-effect transistors *Adv. Mater.* 21 1473–86 (https://doi.org/10.1002/adma.200802733)

Kymissis I 2009 *Organic Field Effect Transistors. Theory, Fabrication and Characterization* (New York: Springer)

Fahlman M, Fabiano S, Gueskine V, Simon D, Berggren M, and Crispin X 2019 Interfaces in organic electronics *Nat. Rev. Mater.* 4627–50 (https://doi.org/10.1038/s41578-019-0127-y)

Problems

8.1 What is the fundamental difference between an OLED and an OSC?

8.2 An OSC with dimensions of 10×10 cm^2 has an open-circuit voltage of 1 V and a short-circuit current of 1 A under the standard illumination of 1 Sun.

- (a) Calculate the efficiency of the OSC.
- (b) Calculate the area of this OSC needed to supply a home with an average monthly consumption of 1000 kWh, assuming an average of 8 h of sun per day at the standard illumination of 1 Sun, and that you have the (battery) means to store the energy at 90% efficiency.

8.3 What are the main differences between an OFET and an OECT?

8.4 An OECT with a channel length of 10 μm, a width of 10 μm, and a thickness of 200 nm was manufactured using PEDOT:PSS with a mobility of 1 cm^2 V^{-1} s^{-1}. The measured transconductance was 100 μS at a gate voltage of 0.2 V in the saturation regime.

- (a) Calculate the volumic and surface capacitances, considering the threshold voltage to be negligible.
- (b) How can the transconductance be improved?

References

[1] Rogers J, Malliaras G and Someya T 2018 *Sci. Adv.* **4** eaav1889
[2] Tang C W and Van Slyke S A 1987 *Appl. Phys. Lett.* **51** 913–5
[3] Burroughes J H, Bradley D D C, Brown A R, Marks R N, Mackay K, Friend R H, Burns P L and Holmes A B 1990 *Nature* **347** 539–41
[4] Baldo M A, Thompson M E and Forrest S R 2000 *Nature* **403** 750–3
[5] Chiang C J, Kimyonok A, Etherington M K, Griffiths G C, Jankus V, Turksoy F and Monkman A P 2013 *Adv. Funct. Mater.* **23** 739–46
[6] Hu D, Yao L, Yang B and Ma Y 2015 *Philos. Trans. R. Soc.* A **373** 20140318
[7] Peng Q, Obolda A, Zhang M and Li F 2015 *Angew. Chem. Int. Ed.* **54** 7091
[8] Ai X, Evans E W, Dong S, Gillett A J, Guo H, Chen Y, Hele T J H, Friend R H and Li F 2018 *Nature* **563** 536–40
[9] Baldo M A, O'Brien D, You Y, Shoustikov A, Sibley S, Thompson M and Forrest S 1998 *Nature* **395** 151–4
[10] Baldo M A, Lamansky S, Burrows P E, Thompson M E and Forrest S R 1999 *Appl. Phys. Lett.* **75** 4–6
[11] Kido J, Kimura M and Nagai K 1995 *Science* **267** 1332–4
[12] Chapin D M, Fuller C S and Pearson G L 1954 *J. Appl. Phys.* **25** 676–7
[13] Tang C W 1986 *Appl. Phys. Lett.* **48** 183–5
[14] Park S H, Roy A, Beaupré S, Cho S, Coates N, Moon J S, Moses D, Leclerc M, Lee K and Heeger A J 2009 *Nat. Photonics* **3** 297–302
[15] Shockley W and Queisser H J 1961 *J. Appl. Phys.* **32** 510–9
[16] Giebink N C, Wiederrecht G P, Wasielewski M R and Forrest S R 2011 *Phys. Rev.* B **83** 195326
[17] Elumalai N K and Uddin A 2016 *Energy Environ. Sci.* **9** 391–410
[18] Grancini G, Maiuri M, Fazzi D, Petrozza A, Egelhaaf H J, Brida D, Cerullo G and Lanzani G 2013 *Nat. Mater.* **12** 29–33
[19] Noriega R, Rivnay J, Vandewal K, Koch F P V, Stingelin N, Smith P, Toney M F and Salleo A 2013 *Nat. Mater.* **12** 1038–44
[20] Herrmann D, Niesar S, Scharsich C, Köhler A, Stutzmann M and Riedle E 2011 *J. Am. Chem. Soc.* **133** 18220–33
[21] Etzold F, Howard I A, Mauer R, Meister M, Kim T D, Lee K S, Baek N S and Laquai F 2011 *J. Am. Chem. Soc.* **133** 9469–79
[22] Jailaubekov A E *et al* 2013 *Nat. Mater.* **12** 66–73
[23] Schulze M, Haensel M and Tegeder P 2014 *J. Phys. Chem.* C **118** 28527–34
[24] Gautam B R *et al* 2016 *Adv. Energy Mater* **6** 1301032
[25] Yin Z, Wei J and Zheng Q 2016 *Adv. Sci.* **3** 1500362
[26] Cheyns D, Vasseur K, Rolin C, Genoe J, Poortmans J and Heremans P 2008 *Nanotechnology* **19** 424016
[27] Farinhas J, Ferreira Q, Paolo R, Alcácer L, Morgado J and Charas A 2011 *J. Mater. Chem.* **21** 12511–9
[28] Liu Q *et al* 2020 *Sci. Bull* **65** 272–5
[29] He K, Kumar P, Yuan Y and Li Y 2021 *Mater. Adv.* **2** 115–45
[30] Liu W, Xu X, Yuan J, Leclerc M, Zou Y and Li Y 2021 *ACS Energy Lett.* **6** 598–608
[31] Xue R, Zhang J, Li Y and Li Y 2018 *Small* **14** 1801793
[32] Chang Y, Zhu X, Lu K and Wei Z 2021 *J. Mater. Chem.* A **9** 3125–50

[33] Sirringhaus H, Friend R H, Li X C, Moratti S C, Holmes A B and Feeder N 1997 *Appl. Phys. Lett.* **71** 3871–3
[34] Chua L L, Zaumseil J, Chang J F, Ou E C W, Ho P K H, Sirringhaus H and Friend R H 2005 *Nature* **434** 194–9
[35] Muccini M and Toffanin S 2016 *Organic Light-emitting Transistors: Towards the Next Generation Display Technology* (Hoboken, NJ: Wiley)
[36] Hepp A, Heil H, Weise W, Ahles M, Schmechel R and von Seggern H 2003 *Phys. Rev. Lett.* **91** 157406
[37] Zaumseil J, Friend R H and Sirringhaus H 2006 *Nat. Mater.* **5** 69–74
[38] Capelli R, Toffanin S, Generali G, Usta H, Facchetti A and Muccini M 2010 *Nat. Mater.* **9** 496–503
[39] Chen H, Xing X, Miao J, Zhao C, Zhu M, Bai J, He Y and Meng H 2020 *Adv. Opt. Mater.* **8** 1901651
[40] Chaudhry M U, Muhieddine K, Wawrzinek R, Sobus J, Tandy K, Lo S C and Namdas E B 2020 *Adv. Funct. Mater.* **30** 1905282
[41] Qin Z, Gao H, Dong H and Hu W 2021 *Adv. Mater.* **33** 2007149 (https://doi.org/10.1002/adma.202007149)
[42] White H S, Kittlesen G P and Wrighton M S 1984 *J. Am. Chem. Soc.* **106** 5375–7
[43] Nilsson D, Robinson N, Berggren M and Forchheimer R 2005 *Adv. Mat.* **17** 353–8
[44] Afonso M, Morgado J and Alcácer L 2016 *J. Appl. Phys.* **120** 165502
[45] Bernards D A and Malliaras G G 2007 *Adv. Funct. Mater.* **17** 3538–44
[46] Rivnay J *et al* 2015 *Sci. Adv.* **1** e1400251
[47] Paudel P R, Tropp J, Kaphle V, Azoulay J D and Lüssem B 2021 *J. Mater. Chem.* C **9** 9761–90
[48] Nissa J, Janson P, Simon D T and Berggren M 2021 *Appl. Phys. Lett.* **118** 053301
[49] Berggren M and Richter-Dahlfors A 2007 *Adv. Mater.* **19** 3201–13
[50] Malliaras G G 2013 *Biochim. Biophys. Acta Gen. Subj.* **1830** 4286–7
[51] Rivnay J, Owens R M and Malliaras G G 2014 *Chem. Mater.* **26** 679–85
[52] Rivnay J, Inal S, Salleo A, Owens R M, Berggren M and Malliaras G G 2018 *Nat. Rev. Mater* **3** 17086
[53] Sophocleous M, Contat-Rodrigo L, García-Breijo E and Georgiou J 2021 *IEEE Sens. J.* **21** 3977–4006

IOP Publishing

The Physics of Organic Electronics
From molecules to crystals and polymers
Luís Alcácer

Chapter 9

Molecular-scale electronic devices

There's plenty of room at the bottom: an invitation to enter a new field of physics.
— Richard Feynman
(American Physical Society meeting, 1959 [1])

This chapter addresses the topic of molecular-scale electronics. The Langmuir–Blodgett (LB) and self-assembled monolayer (SAM) techniques will be introduced, followed by an analysis of transport properties at the nanoscale. Examples of molecular wires, molecular rectifiers, and transistors will be given.

9.1 Introduction

As IBM unveils 2 nm chip technology (ca. 50 billion transistors on a chip, May 2021), we may wonder when Moore's law [2] (stating that the number of transistors on a microchip doubles every two years) will reach its limit. But, as Feynman said in 1959, 'There is plenty of room at the bottom', and the technological trend of adding more and more functionality per cm^3 does not seem to slow down. After all, the human brain has about 2×10^{11} neurons linked to one another via ca. 10^{14} synapses in about 1200 cm^3.

Most organic molecules are larger than 2 nanometers, so it seems that the unimolecular transistor will not overtake Moore's law. However, molecular-scale devices are of great scientific and technological interest, not in competition with the well-established silicon technology, but as models for theoretical development and to perform specific functions by taking advantage of their peculiar potentialities, such as self-assembly and biocompatibility, which are particularly important for the life sciences and healthcare applications.

The simplest molecular (or atomic) device that we can imagine is a switch. It must have two distinct stable and addressable states: an 'ON' state and an 'OFF' state.

doi:10.1088/978-0-7503-3347-4ch9

Then we can think of memories, wires, diodes, transistors, and so on, eventually forming molecular complex systems that can self-assemble and replicate.

Transport at the nanoscale, i.e. the way in which electrons move through a molecule, a molecular wire, or a molecular junction is one of the most important scientific issues, but one of the most difficult practical problems is how to attach electrodes to a single molecule. These problems started to be of concern at the end of the 1960s and the beginning of 1970s with the increasing interest in organic metals and semiconductors. An example of these early studies is the report on tunnelling through fatty-acid salt monolayers, which was achieved using the LB technique by Hans Kuhn and co-workers [3]. However, it was the innovative concept of a single-molecule rectifier envisioned by Aviram and Ratner in 1974 that opened the way to further developments [4]. Their molecular rectifier consisted of a donor π-system and an acceptor π-system, separated by a σ-bonded tunnelling bridge (a D-σ-A molecule).

The ability to 'see', 'touch', and 'manipulate' atoms and molecules arrived in 1981 with the invention of the scanning tunnelling microscope (STM) by Binnig and Rohrer [5, 6], which led to the atomic force microscope (AFM) [7] and several other instruments that are now essential to many fields in nanoscience and nanotechnology.

Rectifiers of the D-σ-A (and D-π-A) types were reported from 1990 onward, using Langmuir–Blodgett molecular monolayers sandwiched between two identical or dissimilar metal electrodes (e.g. Au, Al) [8–14], to mention but a few. LB or Langmuir–Schaefer (LS), films are nanostructured systems of molecules transferred to a substrate from a liquid/gas interface containing a monolayer of amphiphilic molecules.

Studies of molecular rectifiers have also reported single-molecule contacts. For example, Elbing *et al* designed and synthesized a molecule consisting of two coupled π-systems of the D–A type, which behaved as a *single-molecule diode* [15]. Contact between the molecules and the electrodes was achieved using the *mechanically controlled break junction* (MCBJ) technique. This technique, introduced by Moreland and Ekin to study electron tunnelling into superconducting filaments [16], was further developed by Muller and van Ruitenbeek to investigate atomic point contacts [17], and was used for the first time by Mark Reed and co-workers to measure the conductance of a molecular junction [18]. The MCBJ technique consists of forming nanocontacts by pulling a ductile metal wire until it narrows down to the atomic scale and eventually breaks. When it breaks in the presence of molecules (achieved, for example, by adding a droplet of a solution of the molecules to the top of the MCBJ) one or more molecules may eventually connect with the contacts and form a molecular junction. MCBJs can also be very accurately fabricated by lithography, as used for example by Reichert *et al* in their study of a self-assembled metal–molecule–metal junction [19]. Molecules with structures of the type P–S–π–σ–π–S', where P is a protecting group to be removed (e.g. acetyl) and S and S' (e.g. –SH, -NH_2) are end groups that chemically bind to the metal electrodes, have been used in SAMs on Au wires for *I–V* measurements of metal–molecule–metal junctions. In this system, the current rectification could be chemically tuned by controlling the strength of the interaction with the metal contacts [20]. SAMs on gold-coated substrates or highly oriented pyrolytic graphite (HOPG) have been used, for example, by Ashwell *et al* [21–23] to determine *I–V* characteristics by scanning tunnelling spectroscopy, using Au and PtIr tips. SAMs on HOPG are currently commonly used to obtain molecular-scale devices.

9.2 Transport at the nanoscale

At the nanoscale, the transport of charge (and matter) reveals its quantum nature through *conductance*, which changes only in discrete steps and whose size is determined by Planck's constant. In chapter 2, we introduced the resistance quantum, $R_K = h/e^2$ (von Klitzing constant) in the context of the quantum Hall effect. The corresponding conductance quantum is $G_K = e^2/h$.

Electrical conduction in macroscopic systems is described by Ohm's law, and, in the simplest electron gas models, it is conditioned by scattering. In atomic- or molecular-sized conductors, such as nanowires, the transport properties are determined by quantum coherence, which depends on the length scales involved. An important length scale is the *phase-coherence length*, L_ϕ, which is the distance over which the phase of the electron wave function is preserved, and which for a molecular-sized conductor is usually larger than its length, L. When the mean free path, ℓ (the mean distance between successive collisions), is much shorter than the length of the conductor, $\ell \ll L$, the conductance is diffusive, with many scattering events between electrodes, but for $\ell > L$, there are no changes in momentum (no scattering) and the transport is *ballistic*.

The most popular theory to describe coherent transport at the nanoscale (e.g. in atomic or molecular wires) is based on Landauer's idea, who viewed conductance as the *transmission probability* of carriers moving from one terminal to another [24]. Here, we will give a simplified derivation of Landauer's formula (see [25]) that considers the ballistic and adiabatic transport between two metal contacts (reservoirs) with chemical potentials (Fermi levels) μ_L (left) and μ_R (right) through a path with a 'conducting constriction' (also known as a 'quantum point contact') of length L and width W, which are of the order of the Fermi wavelength, λ_F (see figure 9.1(a)).

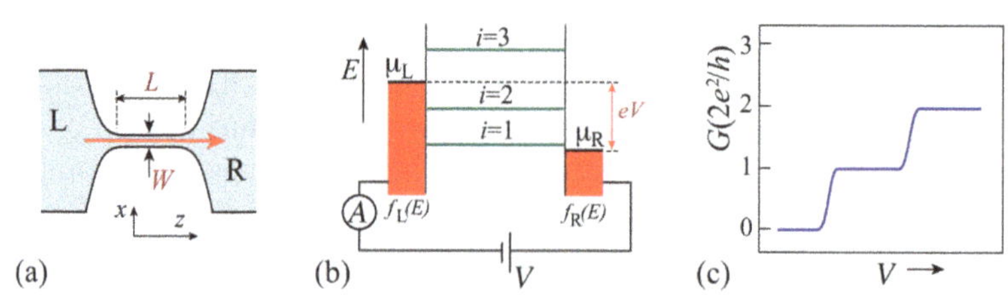

Figure 9.1. (a) Conducting constriction in a 2D electron gas linking two reservoirs, L (left) and R (right). (b) Energy-level diagram of the system. μ_L and μ_R are the chemical potentials of the reservoirs, and $f_j(E)$ are the Fermi–Dirac distributions at $T = 0$. The horizontal lines represent the energy levels of the quantum modes (channels) formed in the transverse direction. The electron motion is confined to the width of the constriction, W, but electrons are free to move along the conductor (in the z direction). (c) Typical curve of the conductance in units of G_0 showing the $i = 1, 2$ plateaux due to the contributing states driven by $\mu_L - \mu_R = eV$.

Note that the quantum behaviour of the moving electrons is determined by the width, W, of the constriction, which allows the passage of just two electrons through each open (energy) channel (due to the Pauli exclusion principle), depending on their Fermi wavelengths, since they are moving as waves. The length of the constriction only determines its ballistic behaviour.

Assuming a constant potential along z and that electron motion is confined in the transverse direction (x), the electron energy is

$$E = E_i + \frac{\hbar^2 k_z^2}{2m}$$

where E_i are the transverse energy levels (modes, or channels), which, for simplicity, we assume to be the solutions of the Schrödinger equation for an electron in a 1D box of dimension W (figure 9.1(b)), with eigenvalues $E = \hbar^2 k_x^2/2m$, wave functions $\psi(x) = A \sin kx$ and $k_x = i\pi/W$ ($i = 1, 2, 3\ldots$). Setting $W = 1$, the density of states for the ith channel is

$$g_i(E) = \frac{\mathrm{d}N}{\mathrm{d}E} = \frac{\mathrm{d}N}{\mathrm{d}k_x}\frac{\mathrm{d}k_x}{\mathrm{d}E} = \frac{1}{\pi}\frac{1}{\hbar v_g(E)} = \frac{2}{hv_g(E)}, \quad \text{with } v_g(E) = \frac{\mathrm{d}\omega}{\mathrm{d}k_x} = \frac{1}{\hbar}\frac{\mathrm{d}E}{\mathrm{d}k_x}.$$

Here, v_g is the group velocity. The current (which only involves electrons near E_F) in the ith channel can now be written as

$$I_i = e^2\, Vg_i(E_F)\, v_{g_i}(E_F) = \frac{2e^2}{h}V, \qquad \text{if} \quad E_i < E_F \quad (E_F \equiv \mu). \tag{9.1}$$

Note that each channel is spin degenerate and would contribute a factor of two in equation (9.1), but this factor is cancelled out since we are interested in electrons with $k > 0$ only (moving from left to right). The conductance of the ith mode can thus be written as:

$$G_i = I_i/V = G_0 = 2e^2/h = 7.748\,091\,73 \times 10^{-5}\,\mathrm{S},$$

which is independent of V and the same for every mode—it is the *conductance quantum*. This value is considered *exact* by CODATA. We can then write the total conductance as

$$G = \frac{1}{V}\sum_i I_i = G_0 \sum_i \Theta(E_F - E_i) = nG_0,$$

$$\Theta(E_F - E_i) = \begin{cases} 0 & \text{if } E_F < E_i \\ 1 & \text{if } E_F > E_i \end{cases} \quad \mu_L > E_F > \mu_R,$$

where μ_L and μ_R are the chemical potentials of the left and right reservoirs, respectively, Θ is the Heaviside step function, and n is the number of modes available between the chemical potentials, each contributing a conductance quantum. This equation shows that when the voltage is changed, the number of channels is changed (the chemical potential is changed in a controlled way by the applied voltage) and the conductance changes stepwise in units of $G_0 = 2e^2/h$, as shown in figure 9.1(c), where $i = 1, 2$ are the allowed plateaux.

In the context of Landauer's idea, the conductance of a *quantum point contact* is

$$G = \frac{2e^2}{h} \sum_{i=1}^{n} T_i. \tag{9.2}$$

This is Landauer's formula, where T_i is the transmission probability of the ith channel. It applies to any situation in which the width of the contact becomes so small that we cannot ignore the wave nature of the moving electrons, i.e. of the order of the Fermi wavelength, λ_F. It applies to scanning tunnelling microscopy experiments, where the constriction is a single atom or molecule connecting the tip of the microscope to the sample surface, and to nanowires, heterojunctions, etc. usually at very low temperatures (ideally at $T = 0$). Gold is the archetypal metal, in which the first plateau (at $1G_0$) extends up to 2 V, corresponding to current densities of the order of 2×10^{15} A m^{-2} [26]. In our simplified derivation of the Landauer formula, we introduced the concept of channels as the energy levels of the particle in a box, but in any real material, we should use the levels that contribute to the energy bands at the Fermi level, E_F. In gold atoms, this is the energy level of the 6*s* orbital, for which the transmission coefficient is $T(E_F) \approx 1$, implying a conductance per atom of $G = 1G_0$ (the macroscopic conductivity is 45.5×10^6 S m^{-1}). In Al, for example, three channels contribute to the conductance, an sp_z with a high transmission coefficient and two with $p_x p_y$ character and low transmission coefficients. The conductance per atom is $G = 0.95G_0$; the low value of the transmission coefficient is attributed to interference effects.

Conductance quantization in steps of $2e^2/h$ was first observed in 1987 by van Vees *et al* [27] and Wharam *et al* [28] in 2D electron gases of GaAs–AlGaAs heterostructures by controlling the width of a narrow channel with a gate (or split gate) on top of the heterojunction.

Note how fundamental the conductance quantum is; this is even more evident in the ballistic transport of neutral atoms in which G_0 is just $1/h$, since neutral atoms have no charge [29]. Equation (9.2) is valid for $T = 0$. For $T > 0$, one can write a version of Landauer's formula for the current that introduces the Fermi–Dirac distributions:

$$\begin{aligned} I(V) &= \frac{2e}{h} \int_{-\infty}^{\infty} T(E)[f_L(E) - f_R(E)]\mathrm{d}E, \\ f_i &= \frac{1}{1 + \exp[(E - \mu_i)/k_B T]} \quad (i = \mathrm{L, R}). \end{aligned} \tag{9.3}$$

The transmission coefficients of (9.2), T_i, or $T(E)$ in (9.3), can be calculated by many different quantum mechanical methods, including tunnelling through potential barriers.

We will now derive the simple Schrödinger equation for a square tunnel barrier. To describe the tunnelling of a particle through a potential barrier in a 1D system, we must solve the Schrödinger equation for three distinct regions, I, II, and III (see figure 9.2(a)).

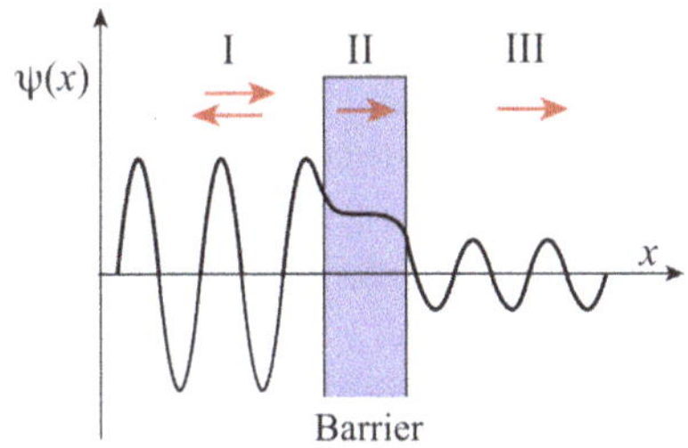

Figure 9.2. Tunnelling effect. The wave function (plane wave) impinging on a potential barrier is partially reflected with a probability amplitude of r and partially transmitted with a probability of $T = |C|^2/|A|^2$ (see text).

In region I, the Schrödinger equation for a free-moving particle is given by

$$-\frac{\hbar^2}{2m}\frac{\mathrm{d}}{\mathrm{d}x^2}\psi_{\mathrm{I}} = E\,\psi_{\mathrm{I}}$$

whose solutions are of the form

$$\psi_{\mathrm{I}} = A\mathrm{e}^{ikx} + B\mathrm{e}^{-ikx}, \qquad \text{with} \quad k^2 = 2mE/\hbar^2.$$

The first and second terms represent the incident and reflected wave, respectively. In region II, the potential barrier ϕ is included in the Schrödinger equation, giving the solution

$$\psi_{\mathrm{II}} = A'\mathrm{e}^{\kappa x} + B'\mathrm{e}^{-\kappa x}, \qquad \text{with} \quad \kappa^2 = 2m(\phi - E)/\hbar^2.$$

In this region, the exponential $\exp(-\kappa x)$ implies a decrease in the wave function amplitude as the particle crosses the barrier. The exponential $\exp(\kappa x)$ appears to have no physical meaning, and we will ignore it. In region III, $\phi = 0$, and the Schrödinger equation is identical to that of region I, but only the forward solution has a physical meaning:

$$\psi_{\mathrm{III}} = C\mathrm{e}^{ikx}$$

and represents the transmitted wave. The probability that the particle (e.g. electron) crosses the barrier is given by the transmission coefficient $T = |C|^2/|A|^2$. An elaborate calculation of the tunnelling current could be made by considering that it should be proportional to the transmission coefficient and depend on the height, ϕ, and thickness, d, of the barrier. It is common to use a simplified expression of the form [30]

$$I = I_0\exp(-\beta d), \qquad \beta = 2\sqrt{2m^*\phi/\hbar^2}, \tag{9.4}$$

where β is the tunnelling decay coefficient, I_0 is the zero-length current, and m^* is the effective mass of the electron. This equation accurately describes coherent non-resonant tunnelling at low bias and can be used in combination with Landauer's idea in order to relate transmission probability and conductance.

In STM experiments, different conduction mechanisms have been observed when the tip of the STM approaches and contacts the atoms or molecules adsorbed on the surface of the substrate, as described by Kröger *et al* [25]. As the tip, initially at ca. 5 Å from the surface, approaches the atoms on the surface, the conductance vs displacement is linear on a log scale, indicating that the current increases exponentially as the distance of the tip to the surface decreases. This is typical of a tunnelling regime described by equation (9.4). When the displacement reaches a critical value, a discontinuous transition from tunnelling to contact occurs (due to relaxation of the tip and surface structures) and the conductance rises to a plateau at the conductance quantum, $G = G_0 = 2e^2/h$, characteristic of ballistic electron transport through a single atom. Interestingly enough, when the tip is retracted after the tunnelling–contact transition, a single adatom is found in ca. 70%–80% of cases. This is a strong indication that the tip and the surface were connected by a single atom, which has also been confirmed experimentally and by simulations [31].

Different kinds of tunnelling can occur in nanosystems, such as single-molecule junctions. In coherent tunnelling, information about the phase of the wave function is preserved and there are no inelastic interactions within the molecular junction. There are, however, several possible tunnelling transport mechanisms, some of which depend on the shape of the potential barrier. Figure 9.3 reproduces results from Ikuno *et al* [32] showing how the shape of the potential barrier can determine the transport mechanism in an STM experiment involving a molecular junction between a metallic tip (e.g. PtIr) and a HOPG substrate.

As shown in figures 9.3(a) and (b) there are two tunnelling mechanisms: direct tunnelling, which occurs when the height of the potential barrier is larger than the applied voltage, $\phi > eV$; and the so-called Fowler–Nordheim (FN) tunnelling [33], which occurs if the potential barrier is smaller than the applied voltage, $\phi < eV$; these mechanisms lead to different types of I–V curve.

The correlation between the tunnelling current, I, and the bias voltage, V, is given by

$$I \propto V \exp\left(-\frac{2d\sqrt{2m^*\phi}}{\hbar}\right) \qquad (V<V_{\text{trans}}) \quad \text{Direct tunnelling}$$

$$I \propto V^2 \exp\left(-\frac{4d\sqrt{2m^*\phi^3}}{3\hbar eV}\right) \qquad (V>V_{\text{trans}}) \quad \text{FN tunnelling,}$$

where m^* is the effective mass of the electrons and d is the barrier thickness.

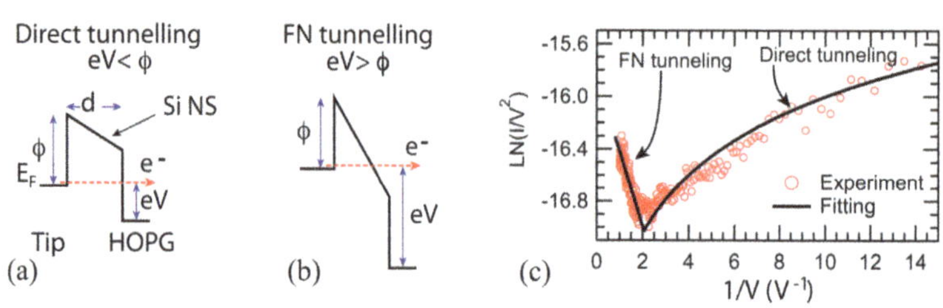

Figure 9.3. (a) and (b) Diagrams of the barrier shapes for direct tunnelling ($eV < \phi$) and Fowler–Nordheim (FN) tunnelling ($eV > \phi$), respectively. (c) Fowler–Nordheim plot of *n*-decylamine-functionalized Si nanosheets (Si NS) and fitting curves (solid line). Reprinted from [32] by permission of AIP Publishing.

The transition voltage from direct to FN tunnelling, V_{trans}, is defined as $V_{trans} = \phi/e$, and can be determined from the FN plot, which is the plot of $\ln(I/V^2)$ vs $(1/V)$:

$$\ln(I/V^2) \propto \ln(1/V) \quad (V < V_{trans}) \quad \text{Direct tunnelling} \tag{9.5}$$

$$\ln(I/V^2) \propto -(1/V) \quad (V > V_{trans}) \quad \text{FN tunnelling.} \tag{9.6}$$

Figure 9.3(c) shows an FN plot and the fitting curves for *n*-decylamine-functionalized Si NS, showing the two distinct types of tunnelling regime depending on the applied voltage, as reported by Ikuno *et al* [32]. In molecular junctions, the potential barrier height, ϕ, is frequently taken to be the approximate difference between the Fermi energy, E_F, of the electrodes and the energy of the nearest frontier molecular orbital, E_{HOMO}, or E_{LUMO}.

Other temperature- and length-dependent transport mechanisms are common in molecular junctions, such as hopping conduction and thermionic conduction, which are strongly temperature dependent. Space-charge limited conduction also takes place, which is only length dependent:

$$\begin{aligned} &I \sim Vd\exp(-\phi/k_BT) && \text{Hopping} \\ &I \sim T^2\exp\left[-(\phi - e\sqrt{eV/4\pi\varepsilon d})/k_BT\right] && \text{Thermionic emission} \\ &I \sim V^2/d^3 && \text{Space–charge limited,} \end{aligned} \tag{9.7}$$

where ϕ is the potential barrier height, d is the molecular length, and ε is the permittivity ($\varepsilon = \varepsilon_r\varepsilon_0$, with ε_r, the relative permittivity, and ε_0, the vacuum permittivity). Thermionic emission is a mechanism which occurs when electrons are excited and go over the potential barrier.

9.2.1 Transport through single-molecule junctions

Single-molecule junctions are devices consisting of a single molecule electrically connected to two electrodes, one at each end. The electrical properties of molecular junctions, namely the conductance at a given voltage, $G = I/V$, and the current–voltage characteristics, *I*–*V*, have been measured by several methods, including MCBJs and variants of scanning tunnelling microscopy (STM) and atomic force microscopy (AFM). Current-sensing AFM (CSAFM), conductive probe AFM (CPAFM) [34], and STM break junctions (STMBJs) [35] are the most popular approaches. The CSAFM method is easy to use, but the results are affected by the large size of the tip radius, implying contact with several molecules at the same time. The STMBJ method is more reliable for measurements of single-molecule junctions, since the STM tip is chemically bonded to the molecule. To determine the conductance of a single molecule, large numbers of measurements are statistically analysed using conductance histograms, which show peaks at integer multiples of $\approx G_0$ in the case of Au–Au contacts whose thickness is that of a single atom, or small fractions of G_0 (typically two orders of magnitude less than G_0) in the case of molecular junctions, as shown in figure 9.4, for Au–BDT–Au (BDT is benzene-1,4-dithiol) junctions.

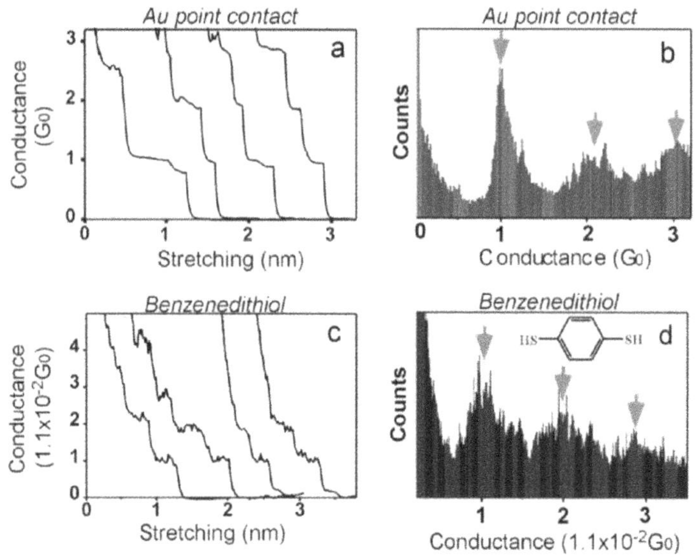

Figure 9.4. By repeatedly pulling an Au STM tip from an Au surface in the presence of a few BDT molecules, the conductance first decreases by steps of $G_0 = 2e^2/h$ (a), leading to peaks near $1G_0$, $2G_0$, and $3G_0$, in the conduction histogram (b). After further pulling, the gold contact is broken, and a new sequence of steps eventually appears in the conductance curves (c), leading to peaks in the corresponding histogram (d) at 0.011, 0.022, and 0.033 G_0 due to the conductance of one, two, and three BDT molecular junctions, respectively. The conductance curves are displaced along x for the sake of clarity. Reprinted with permission from [36]. Copyright 2004, American Chemical Society.

In the left part of figures 9.4(a) and (c), the conductance is plotted against the stretching length (in nm) of the contact (conductance curve) for an Au–Au contact (a) and a single-molecule junction (c). The steps in the conductance curve are due to the formation of molecular junctions. The histograms on the right are plots of the number of conductance curves vs the conductance in units of G_0 for the respective conductance curve sets.

In molecular junctions, tunnelling is the main transport process; it follows equation (9.4) for short molecules (a few nm long) and the hopping equation (9.7) for long molecules.

As discussed above, a single gold atom contact has a conductance of $\approx 1G_0$. It is interesting to note that there are examples of single-molecule junctions with equally high conductances. A curious one is that of a hydrogen molecule linked to Pt atoms (a single-channel Pt–H–H–Pt junction). Using the MCBJ technique at 4.2 K, Smit *et al* obtained a conductance close to $1G_0$. They simulated the H_2-bridge configuration using density functional theory (DFT) and calculated the value $G = 0.9G_0$ [37].

Current–voltage characteristics of single-molecule junctions

The *I–V* characteristics of a single-molecule junction are determined by the positions of the molecular levels, E_i, relative to E_F, the nature of the molecular levels, and the tunnel coupling constants Γ_L and Γ_R. There are essentially two transport regimes:

direct tunnelling if $eV \ll |E_i - E_F|$ for any molecular level i (i.e. there are no molecular energy levels available to serve as channels inside the bias window, eV), and resonant transport through the discrete molecular levels when $eV \approx 2|E_i - E_F|$ for at least one level. This behaviour was described in the seminal work of Mark Reed [18], and in many subsequent studies, such as that of Lörtscher *et al*, who also used the MCBJ method and a detailed statistical analysis, particularly for Au–BDT–Au junctions [38].

Figure 9.5 represents an Au–BDT–Au junction, in which the sulphur atoms bind to the gold atoms in (a); a typical energy-level diagram of a junction (not necessarily Au–BDT–Au) showing the chemical potentials, μ_L and μ_R, of the electrodes and the energy levels of the π-type orbitals of the molecule, the LUMO being the most suitable channel for the current, in (b). In (c) we show the typical shape of an I–V curve for a molecular junction: a stepwise (or S-like) shape, in which the current (of the order of μA or nA) increases when the Fermi level approaches a molecular level, reaching a plateau (corresponding to a gap) and increasing again as it approaches an eventual next molecular level, as in (b).

We now use the Landauer formula (9.3), using the transmission coefficient, $T(E)$, for a single-channel junction, following, for example, the excellent mini review by Isshiki *et al* [39] that provided examples of the I–V characteristics of single-molecule junctions. For a single-channel junction, the transmission coefficient $T(E)$ of equation (9.3) is a Lorentzian-type distribution of the form

$$T(E) = \frac{4\,\Gamma_L \Gamma_R}{(\Gamma_L + \Gamma_R)^2 + (E - E_0)^2}, \tag{9.8}$$

where Γ_L and Γ_R are the tunnel coupling constants of the electrodes (assumed to be independent of the energy), and E_0 is the molecular energy level relative to the Fermi level, E_F. The transmission coefficient (9.8) can be used in the Landauer formula (9.3), making $\Gamma = \Gamma_L + \Gamma_R$ and $\alpha = \Gamma_L/\Gamma$ to give the current going through the junction when a voltage, V, is applied:

$$I = \frac{8e}{h}\alpha(1 - \alpha)\Gamma\left\{\arctan\left[\frac{\alpha eV - E_0}{\Gamma}\right] + \arctan\left[\frac{(1 - \alpha)eV + E_0}{\Gamma}\right]\right\}. \tag{9.9}$$

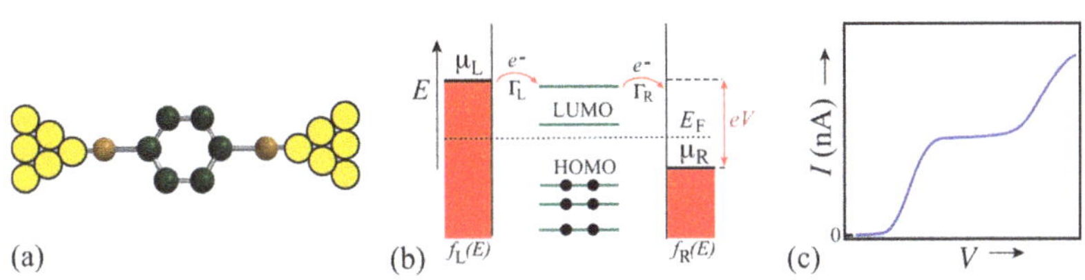

Figure 9.5. (a) Schematic representation of a molecular junction, where a conjugated molecule is connected to Au contacts via a thiol link (–S–Au). (b) Energy-level diagram of the junction, where the levels involved are the molecular levels close to the frontier levels, in particular the LUMO and the LUMO+1. Γ_L and Γ_R are the tunnel coupling constants of the electrodes, and the $f_i(E)$ are the Fermi–Dirac distributions at $T = 0$. (c) Typical I–V characteristic for the transport through the molecular levels lying between μ_L and μ_R.

The current is an arctangent-type function that varies stepwise with the applied bias, as schematically depicted in figure 9.5(c) for weak coupling and low temperatures. The *I–V* characteristics have an S-like shape that increases when the chemical potentials of the electrodes approach the molecular energy level E_0 (the resonant condition). By fitting equation (9.9) to the measured *I–V* curves, we can determine the values of Γ and E_0 which, according to Isshiki *et al*, were $\Gamma = 135$ meV and $E_0 = 266$ meV for BDT. This behaviour of $I(V)$ is clearly seen in the results of the *I–V* characteristics of Au–BDT–Au junctions (even at 50 and 250 K) as reported, for example, by Lörtscher *et al* using the MCBJ method [38]. A peak in a resistance histogram for the applied voltage of 5 mV at 300 K indicates that the resistance of the junction at that temperature is ≈260 MΩ (i.e. $4.96 \times 10^{-5} G_0$). A more recent and detailed study of BDT junctions by Kim *et al* shows that the measurement of molecular junctions is a difficult task due to various factors, including the exact (geometrical) way the S–BDT–S entity makes contact with the gold atoms of the STM tip, leading to different results [40].

9.3 Quantum effects in molecular junctions

Charge transport in molecular junctions is a rich ground for quantum phenomena, such as quantum interference and decoherence, Coulomb blockade, the Kondo effect, Franck–Condon blockade, vibronic effects, and many others.

The observation of quantum interference in molecular junctions is due, in the first place, to the fact that the moving electrons propagate as waves, and when a wave splits and follows two or more different paths that later join, the two or more components of the wave function interfere constructively or destructively according to the difference between their phases. This is a consequence of the superposition principle of quantum mechanics.

The transmission probabilities, T_i, in the conductance (equation 9.2) are, in fact, the squares of the moduli of complex probability amplitudes (the coefficients of the components of the wave function). A more general version of (9.2) could be written as

$$G = G_0 \sum_{ij} |\tau_{ij}|^2$$

where $|\tau_{ij}|^2$ describes the transmission probability for an electron to be transmitted from contact mode i in the left reservoir into contact mode j in the right reservoir [41]. This suggests that the τ_{ij} are elements of a matrix whose diagonal elements form the complete set of transport modes, or *transport eigenchannels*, of the system. T_i are the squares of the moduli of the diagonal elements of the matrix, i.e. $T_i = |\tau_{ii}|^2$. The multichannel conductance equation constitutes the Landauer–Büttiker formalism and is the appropriate way to deal with quantum effects (e.g. interference). This is done using advanced computational methods.

Since in molecular junctions the transmission channels are, in principle, the molecular energy levels (molecular orbitals (MOs)) near the frontier levels, we should write the transmission probabilities in terms of the complex transmission coefficients which describe the amplitude and phase of each individual electron wave. For example, for a molecule with two molecular orbitals, a and b, the total transmission probability should be written as

$$T = |\tau_a + \tau_b|^2 = |\tau_a|^2 + |\tau_b|^2 + \tau_a\tau_b^* + \tau_a^*\tau_b$$
$$T = T_a + T_b + T_{ab}, \qquad T_{ab} = 2\sqrt{T_a T_b}\cos(\theta_a - \theta_b).$$

T_{ab} is the interference term, and θ_a and θ_b are the phases of τ_a and τ_b, respectively. If $\theta_a = \theta_b$, there is constructive interference between the molecular orbitals a and b; if they differ by 180°, there is complete destructive interference. A general expression involving several orbitals can be written as

$$T = \sum_i T_i + \sum_{i>j} T_{ij}.$$

The problem of calculating T_{ij} is complex because the MOs of the isolated molecule are not the eigenchannels of the junction, since the molecule is bonded to the electrodes, and the couplings have to be included in any calculation. This problem has been addressed in many ways, but the most effective involve advanced computational models, which are beyond the scope of this text. It is instructive, however, to look at the results of a recent article by Gunasekaran *et al* [42]. Using an advanced method based on Green's functions, the authors show how the transmission coefficient can be reduced to the elements of a matrix that is converted into an illustrated quantum interference map, in which the non-interfering components are the diagonal elements and the off-diagonal non-zero elements are the interfering ones. The results of several examples of molecular junctions, such as benzene, are displayed in that map to help to visualize the interfering MOs. In the Hückel approximation, the frontier orbitals of the π system of benzene (HOMO and LUMO) are degenerate, but the coupling to the electrodes lifts the degeneracy, implying different transmission coefficients: junctions with contacts in the 'para' geometry show constructive interference at E_F, while contacts in the 'meta' geometry show complete destructive interference at E_F. A Pt–benzene–Pt junction with a conductance of $1G_0$ was reported by Kiguchi *et al* [43], who confirmed the formation of the benzene–Pt chemical bonds not only by conductance measurements but also using other techniques, such as inelastic electron tunnelling spectroscopy (IETS), to test the effect of isotopic substitution on the vibration modes and the dependence of the conductance and transmission channels on the distance between the electrodes. They also performed DFT calculations to study the benzene–Pt junction and calculated the conductance using Green's functions and Landauer's formula. Using shot noise measurements, they decomposed the transmission coefficients and concluded that the current in the junction is carried by three partially transmitting channels which depend on the details of the junction geometry. The calculated stablest distances between electrodes of 3.4, 4.9, and 6.6 Å were assigned to junctions with observed conductances of 0.71, 1.08, and $0.20G_0$, respectively. In principle, the number of conduction channels is equal to the number of C atoms bonded to the atoms of the Pt tip, since each C atom has only one orbital in the benzene π-system and therefore each C atom can only form one π channel.

Interference effects in single-molecule junctions are a major scientific challenge, not only experimentally (due to the various possible geometries) but also theoretically, and there are many approaches to these problems. To get an idea of the main properties of single-molecule junctions beyond transport, in particular, interference phenomena, we recommend the review by Aradhya and Venkataraman [44].

Decoherence effects arise from electronic–vibrational coupling (see, for example [45]). Coulomb blockade consists of the mutual repulsion between electrons inside and outside of the molecular junction and may result in a single-electron transport regime. Single-electron transistors and memories are potential devices based on Coulomb blockade [46, 47]. The Kondo effect, already mentioned in chapter 6, is also observed in molecular junctions as a resistance minimum, which results from exchange interactions between moving electrons with a localized spin (see for example [48, 49]).

For a deeper understanding of transport in nanostructures and the other effects mentioned, we would have to go beyond the scope of this text and resort to computational techniques, entering the field of non-equilibrial transport in many-body quantum systems, and pursuing either transport theory, from the Landauer formula to the non-equilibrium Green's function (NEGF) formalism, molecular quantum dynamics, numerical renormalization group theory, and the Tomonaga–Luttinger liquid theory [50], or in the direction of electronic structure models, from the tight binding method to mean field, DFT, and electron correlation theories. We refer readers interested in the theory of quantum transport in molecular junctions in connection with experimental studies to the paper by Thoss and Evers [51] and the review by Evers *et al* cited in Further reading. We also recommend reviewing the computational method based on DFT and the results for the calculation of the transmission coefficient of the BDT junction reported by Stokbro *et al* [52].

9.4 Examples of molecular electronic devices

The best examples of molecular electronic devices include diodes (rectifiers), switches, memories, wires, field-effect transistors (FETs), electrochemical transistors, and sensors. Other types of molecular-sized devices include, for example, molecular motors and drug delivery devices, which are of particular importance in the life sciences. The single-molecule devices are of particular importance, since they are the building blocks for more complex systems.

Molecular diodes based on the Aviram and Ratner model [4] consist of D-σ-A, D-π-A, or simply D–A junctions with a final gate electrode to adjust the rectification ratio by changing the alignment between the energy levels and the Fermi energy. Their *I*–*V* characteristics follow the theory described above (see figure 9.5). For example, in a simple D–A junction, at zero applied bias, the HOMO level of the donor should be slightly below the electrode's chemical potential and the LUMO level of the acceptor should be slightly above it.

Molecular wires

Molecular wires are molecular chains that can be used as links between elements in nanosystems; they carry electronic charges (electrons and holes), ions, and energy in the form of photoexcitations. There are many examples of molecular wires, some of which are particularly long and good electronic conductors. Conjugated wires with several D-π-A units have been reported, such as one 20.2 nm long with a conductivity of 0.02 S cm^{-1} reported by Choi *et al* [53] and molecular wires formed from large numbers of redox centres incorporated into organic chains by the stepwise complexation of metal centres via self-assembly; the lengths of the molecular wires were as much as 40 nm, in which the contact resistances with the electrodes were the main contributors to the total resistance, the molecular wire itself being highly conductive [54].

Stepwise methods based on molecular self-assembly can provide simple and reproducible ways to prepare supramolecular structures. The scanning tunnelling microscope (STM) is a versatile tool for fabrication and control of the assembly process at the molecular scale. When it is operated at the solid–liquid interface by placing a solvent droplet between the STM tip and the substrate, it is possible to add molecules in order to create organized structures [55]. Ferreira *et al* [56] applied this method to build long molecular wires formed of zinc-octaethylporphyrin (ZnOEP) molecules separated by 4,4-bipyridine molecules. STM was used to assemble the molecules on HOPG. Each molecule was deposited individually, and each step was controlled in real time using molecular-resolution STM images. Wires composed of up to 25 individual molecules arranged in a well-defined sequence and assembled via bonding of the central metal of the porphyrin to the nitrogen atom of the bipyridine were built. The final structures were molecular wires 14 nm long, as determined by DFT, at the B3LYP/6-31G* level. Figure 9.6 shows, on the left, a simulation of the 14 nm wire formed of 13 porphyrin and 12 bipyridine units.

The first STM image on the right is an STM image of the first self-assembled monolayer of porphyrin molecules, while the second is an STM image of the self-assembled layer of porphyrin molecules taken when it was the top layer, i.e. an STM image of the surface of a film with 23 molecular layers (of alternating porphyrin and bipyridine molecules). Note that the twenty-third layer still maintains the high resolution. The arrows shown in both images (indicating that the second is slightly rotated relative to the first) define the nearly 2D hexagonal unit cell with the same dimensions ($a = 1.4 \pm 0.1$ nm, $b = 1.3 \pm 0.1$ nm, and $\gamma = 110 \pm 5\,°$).

The conductance of each wire was measured by scanning tunnelling spectroscopy (STS), and hundreds of *I*–*V* curves were recorded at each growth step. The length dependence of the wires' resistance was analysed to assess the relative importance of the tunnelling and hopping regimes. The results show that these wires are highly conductive, with an ultralow attenuation factor of $\beta = 0.015 \pm 0.006$ Å^{-1} (9.4) for short wires in the tunnelling regime and a transition to the hopping regime at ≈6.5 nm. Using the conductance histogram built from over 2500 individual *I*–*V* curves, the conductance of the 14 nm wires was estimated to be $0.14 \pm 0.01 \times 10^{-5} G_0$, corresponding to a conductivity of ca. 0.02 S cm^{-1} [57].

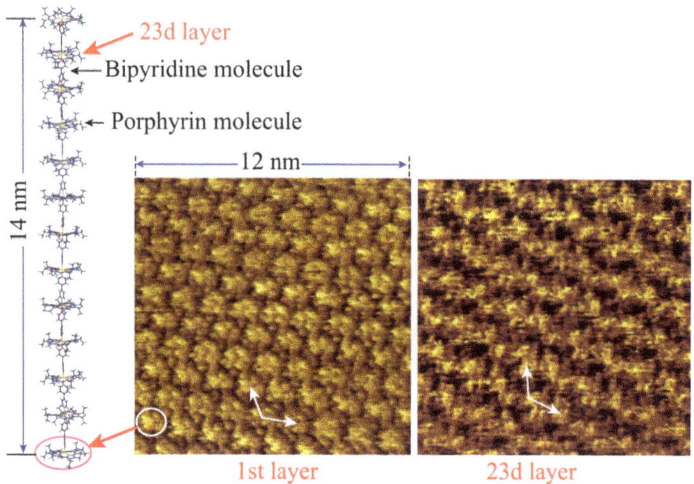

Figure 9.6. Fourteen nanometre conductive molecular wire formed from alternating porphyrin and bipyridine molecules linked to each other, as modelled by DFT B3LYP quantum chemical calculation. On the right, STM images of the first and twenty-third layers, where each repeating motif is a porphyrin molecule anchoring a molecular wire [56, 57]. Note that the twenty-third layer still maintains the high resolution. However, the following layers were affected by noise resulting from molecular vibrations which influence image acquisition. Adapted with permission from [57]. Copyright 2014, American Chemical Society.

Ultralow β values have previously been reported for wires formed of porphyrins linked by axial coordination—a value of $\beta = 0.02$ Å^{-1} was obtained for β-fused oligoporphyrins [58]. The slightly lower result reported by Ferreira *et al* suggests that the conductance is improved if the porphyrins are linked by the metal centre instead of by axial coordination. A review of porphyrins linked by axial coordination has recently been published by Zwick *et al* [59].

Molecular field-effect transistors

A molecular field-effect transistor is a molecular junction in which the current is modulated by an external gate voltage that shifts the frontier molecular energy levels (acting as the junction channels) relative to the chemical potentials of the source and drain electrodes.

The current in a molecular junction can also be modulated by a Coulomb blockade or the Kondo effect, as in the case reported by Park *et al* [46], in which a Co atom was coordinated, in a near octahedral geometry, to two terpyridinyl (tpy) ligands with thiol end groups linked to Au electrodes, with the chemical structures $[Co(tpy\text{-}(CH_2)_5\text{-}SH)_2]^{2+}$ and $[Co(tpy\text{-}SH)_2]^{2+}$. The transistors were prepared by electromigration breakdown of Au wires coated with an SAM of the complex on a SiO_2/doped Si gate. The current flows through the charge states of the Co atom. While the junction behaves as a single-electron transistor in the longer molecule with weak coupling due to Coulomb blockade, in the short strongly coupled molecule, the tunnelling is enhanced by the Kondo effect—a Kondo peak is observed at low temperatures in the $\partial I/\partial V$ vs V curve, which splits in two in a magnetic field.

In a very instructive article, Song *et al* [60] demonstrated direct modulation of a transistor channel current by an external gate voltage that adjusted the relative positions of the frontier molecular orbitals (HOMO and LUMO) with respect to the source and drain. The authors compared the behaviour of transistors using two different molecules: 1,8-octanedithiol (ODT) with an alkyl backbone, as a non-conducting control, and π-conjugated 1,4-benzenedithiol (BDT) as the active transistor. Figure 9.7 illustrates (a) the architecture of the transistors; (b) the $I(V)$ characteristics of the BDT transistor, measured at 4.2 K; (c) the Fowler–Nordheim plots (for the BDT), indicating the voltages at which the transitions from the Fowler–Nordheim regime to the direct tunnelling regime occur.

The transistors were made using the electromigration method, which consists of breaking a gold wire coated with an SAM of the chosen molecules in vacuum at 4.2 K laid down on an Al_2O_3/Al gate. Each transistor consisted of a single molecule chemically bonded to an Au source and drain with the Al gate, in a bottom-gate FET architecture. The ODT transistors exhibited S-shaped $I(V)$ characteristics with drain currents on the nA scale that increased as the gate voltage became more negative (from 0 to −3.3 V). The FN plots, $\ln(I/V^2)$ vs $(1/V)$ (equations 9.5 and 9.6) clearly show the two distinct regimes, i.e. direct and FN tunnelling or field emission, with $V_{trans} = \phi/e$ increasing linearly with V_G in the domain from −3.2 to −1.2 V and a slope of $\alpha = \Delta V_{trans}/\Delta V_G = 0.25$; here, α is the gate efficiency factor, i.e. the efficacy of the molecular orbital adjustment.

The I(V) characteristics of the BDT transistor (figure 9.7(b)) show that its currents are three orders of magnitude higher than those of the ODT devices, i.e. in the μA range for gate voltages in the range of −3 to +3 V; they increase with an increase in negative V_G and decrease with a positive V_G. Figure 9.7(c) shows the FN plots, with arrows indicating the transition, V_{trans}, from FN to direct tunnelling. The reported value of $\alpha = +0.22$ in the BDT transistor is an indication that the main transport channel is the HOMO energy level. The V_{trans} of 1.14 ± 0.04 V for $V_G = 0$ is much

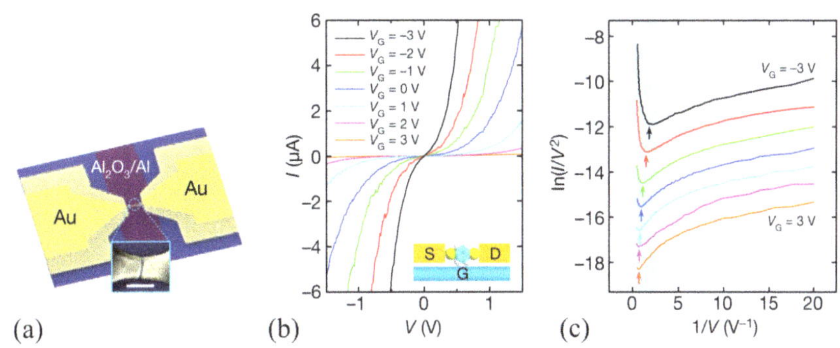

Figure 9.7. (a) Structure of a molecular transistor showing a single molecule, the Au source and drain, and the Al gate on Al_2O_3 (scale bar, 100 nm). (b) $I(V)$ characteristics for different values of V_G of a 1,4-benzenedithiol (BDT) transistor measured at 4.2 K; inset: device schematic. (c) Fowler–Nordheim plots; the arrows indicate the transitions, V_{trans}, from Fowler–Nordheim to direct tunnelling. The plots are vertically offset for clarity. Adapted from [60]. Copyright Springer, 1998, adapted by permission from Springer.

lower than the corresponding value for the ODT transistor because the BDT molecule has a smaller HOMO–LUMO gap, since it is a π-conjugated molecule.

Further reading

Guo X, Xiang D, and Li Y 2020 *Molecular-Scale Electronics: Concept, Fabrication and Applications* (Weinheim: Wiley)

Bâldea I (ed) 2016 *Molecular Electronics. An Experimental and Theoretical Approach* (Singapore: Pan Stanford)

Evers F, Korytár R, Tewari S, and van Ruitenbeek J M 2020 Advances and challenges in single-molecule electron transport *Rev. Mod. Phys.* 92 035001 (https://doi.org/10.1103/RevModPhys.92.035001)

Chen B and Xu K 2019 Single molecule-based electronic devices: a review *Nano* 14 1930007 (https://doi.org/10.1142/S179329201930007X)

Reifenberger R 2016 *Fundamentals of Atomic Force Microscopy: Part I: Foundations* (Singapore: World Scientific)

Problems

9.1
- (a) Perform a dimensional analysis (in terms of SI units) of the conductance quantum $G_0 = 2e^2/h$ for electrons and $G_0 = 1/h$ for neutral atoms.
- (b) The conductance quantum corresponding to the von Klitzing constant is $G_K = e^2/h$, while the value recommended by CODATA is $G_0 = 2e^2/h$. Can you explain the difference, and to what does the CODATA value correspond?

9.2 The conductance of the first plateau (1 G_0) for gold at 2 V corresponds to a current density of the order of 2×10^{15} A m^{-2}.
- (a) Estimate the number of electrons that flow through the point contact per second.
- (b) Assuming that the number of electrons found as the answer to (a) pass through a gold wire one atom thick, find the order of magnitude of the effective radius of the atom and compare it with the value of the 6*s* orbital radius of gold, $r_{6s}(Au) = 156$ pm (from the literature).

9.3 What is self-assembly? Explain how it is possible.

References

[1] Feynman R P 1960 *Eng. Sci.* **23** 22–36 https://calteches.library.caltech.edu/1976/1/1960Bottom.pdf
[2] Moore G E 1965 *Electronics* **38** 114–7
[3] Mann B and Kuhn H 1971 *J. Appl. Phys.* **42** 4398–405
[4] Aviram A and Ratner M A 1974 *Chem. Phys. Lett.* **29** 277–83
[5] Binnig G, Rohrer H, Gerber C and Weibel E 1982 *Physica B+C* **109-10** 2075–7
[6] Binnig G, Rohrer H, Gerber C and Weibel E 1982 *Phys. Rev. Lett.* **49** 57–61
[7] Binnig G, Quate C F and Gerber C 1986 *Phys. Rev. Lett.* **56** 930–3

[8] Ashwell G J, Sambles J R, Martin A S, Parker W G and Szablewski M 1990 *J. Chem. Soc., Chem. Commun.* 1374–6
[9] Martin A S, Sambles J R and Ashwell G J 1993 *Phys. Rev. Lett.* **70** 218–21
[10] Metzger R M *et al* 1997 *J. Am. Chem. Soc.* **119** 10455–66
[11] Metzger R M 1999 *J. Mater. Chem.* **9** 2027–36
[12] Metzger R M, Xu T and Peterson I R 2001 *J. Phys. Chem.* B **105** 7280–90
[13] Xu T, Peterson I R, Lakshmikantham M V and Metzger R M 2001 *Angew. Chem. Int. Ed.* **40** 1749–52
[14] Metzger R M 2006 *Chem. Phys.* **326** 176–87
[15] Elbing M, Ochs R, Koentopp M, Fischer M, von Hänisch C, Weigend F, Evers F, Weber H B and Mayor M 2005 *Proc. Natl. Acad. Sci. U.S.A* **102** 8815–20
[16] Moreland J and Ekin J W 1985 *Appl. Phys. Lett.* **47** 175–7
[17] Muller C J, van Ruitenbeek J M and de Jongh L J 1992 *Phys. Rev. Lett.* **69** 140–3
[18] Reed M A, Zhou C, Muller C J, Burgin T P and Tour J M 1997 *Science* **278** 252
[19] Reichert J, Ochs R, Beckmann D, Weber H B, Mayor M and Löhneysen H v 2002 *Phys. Rev. Lett.* **88** 176804
[20] Kushmerick J G, Whitaker C M, Pollack S K, Schull T L and Shashidhar R 2004 *Nanotechnology* **15** S489–93
[21] Ashwell G J, Hamilton R and High L R H 2003 *J. Mater. Chem.* **13** 1501–3
[22] Ashwell G J, Chwialkowska A and High L R H 2004 *J. Mater. Chem.* **14** 2389–94
[23] Ashwell G J, Chwialkowska A and Hermann High L R 2004 *J. Mater. Chem.* **14** 2848–51
[24] Landauer R 1957 *IBM J. Res. Dev.* **1** 223–31
[25] Kröger J, Néel N and Limot L 2008 *J. Phys.: Condens. Matter* **20** 223001
[26] Agraït N, Yeyati A L and van Ruitenbeek J M 2003 *Phys. Rep.* **377** 81–279
[27] van Wees B J, van Houten H, Beenakker C W J, Williamson J G, Kouwenhoven L P, van der Marel D and Foxon C T 1988 *Phys. Rev. Lett.* **60** 848–50
[28] Wharam D A, Thornton T J, Newbury R, Pepper M, Ahmed H, Frost J E F, Hasko D G, Peacock D C, Ritchie D A and Jones G A C 1988 *J. Phys. C Solid State Phys.* **21** L209–14
[29] Krinner S, Stadler D, Husmann D, Brantut J P and Esslinger T 2015 *Nature* **517** 64–7
[30] Binnig G, Rohrer H, Gerber C and Weibel E 1982 *Appl. Phys. Lett.* **40** 178–80
[31] Limot L, Kröger J, Berndt R, Garcia-Lekue A and Hofer W A 2005 *Phys. Rev. Lett.* **94** 126102
[32] Ikuno T, Okamoto H, Sugiyama Y, Nakano H, Yamada F and Kamiya I 2011 *Appl. Phys. Lett.* **99** 023107
[33] Fowler R H and Nordheim L 1928 *Proc. R. Soc.* A **119** 173–81
[34] Wold D J, Haag R, Rampi M A and Frisbie C D 2002 *J. Phys. Chem.* B **106** 2813–6
[35] Xu B and Tao N J 2003 *Science* **301** 1221–3
[36] Xiao X and Tao N J 2004 *Nano Lett.* **4** 267–71
[37] Smit R H M, Noat Y, Untiedt C, Lang N D, van Hemert M C and van Ruitenbeek J M 2002 *Nature* **419** 906–9
[38] Lörtscher E, Weber H B and Riel H 2007 *Phys. Rev. Lett.* **98** 176807
[39] Isshiki Y, Matsuzawa Y, Fujii S and Kiguchi M 2018 *Micromachines* **9** 67
[40] Kim Y, Pietsch T, Erbe A, Belzig W and Scheer E 2011 *Nano Lett.* **11** 3734–8
[41] Büttiker M, Imry Y, Landauer R and Pinhas S 1985 *Phys. Rev.* B **31** 6207–15
[42] Gunasekaran S, Greenwald J E and Venkataraman L 2020 *Nano Lett.* **20** 2843–8

[43] Kiguchi M, Tal O, Wohlthat S, Pauly F, Krieger M, Djukic D, Cuevas J C and van Ruitenbeek J M 2008 *Phys. Rev. Lett.* **101** 046801
[44] Aradhya S V and Venkataraman L 2013 *Nat. Nanotechnol* **8** 399–410
[45] Ballmann S, Härtle R, Coto P B, Elbing M, Mayor M, Bryce M R, Thoss M and Weber H B 2012 *Phys. Rev. Lett.* **109** 056801
[46] Park J *et al* 2002 *Nature* **417** 722–5
[47] Moth-Poulsen K and Bjørnholm T 2009 *Nat. Nanotechnol* **4** 551–6
[48] Zimbovskaya N A and Pederson M R 2011 *Phys. Rep* **509** 1–87
[49] Rakhmilevitch D, Korytár R, Bagrets A, Evers F and Tal O 2014 *Phys. Rev. Lett.* **113** 236603
[50] Li F 2013 *Bosonization and its application to transport in quantum wires Luttinger Model. The First 50 Years and Some New Directions (Series on Directions in Condensed Matter Physics* vol 20 ed V Mastropietro and D C Mattis (Singapore: World Scientific) pp 185–208
[51] Thoss M and Evers F 2018 *J. Chem. Phys.* **148** 030901
[52] Stokbro K, Taylor J, Brandbyge M, Mozos J L and Ordejón P 2003 *Comput. Mater. Sci.* **27** 151–60
[53] Choi S H and Frisbie C D 2010 *J. Am. Chem. Soc.* **132** 16191–201
[54] Tuccitto N, Ferri V, Cavazzini M, Quici S, Zhavnerko G, Licciardello A and Rampi M A 2009 *Nat. Mater.* **8** 41–6
[55] Ferreira Q, Bragança A M, Moura N M M, Faustino M A F, Alcácer L and Morgado J 2013 *Appl. Surf. Sci.* **273** 220–5
[56] Ferreira Q, Alcácer L and Morgado J 2011 *Nanotechnology* **22** 435604 435604–10
[57] Ferreira Q, Bragança A, Alcácer L and Morgado J 2014 *J. Phys. Chem.* C **118** 7229–34
[58] Sedghi G, Esdaile L J, Anderson H L, Martin S, Bethell D, Higgins S J and Nichols R J 2012 *Adv. Mater.* **24** 653–7
[59] Zwick P, Dulić D, van der Zant H S J and Mayor M 2021 *Nanoscale* **13** 15500–25
[60] Song H, Kim Y, Jang Y H, Jeong H, Reed M A and Lee T 2009 *Nature* **462** 1039–43

IOP Publishing

The Physics of Organic Electronics
From molecules to crystals and polymers
Luís Alcácer

Appendix A

Second quantization

The quantum description of the elementary excitations becomes clearer if one uses the *occupation number formalism* or *second quantization* of *quantum field theory*, which is based on the use of the number of particles that occupy each state in a complete set of states of individual particles. This formalism differs from the first quantization, which is based on the use of the states of individual particles. In this representation, we have to distinguish between bosons and fermions.

Let us start with bosons. Consider the Hamiltonian

$$H = \frac{1}{2}\sum_k \left[P_k^* P_k + \omega_k^2 Q_k^* Q_k\right], \tag{A.1}$$

where Q and P are generalized coordinates and momenta, respectively. This Hamiltonian is formally identical to the sum of the Hamiltonians of the linear harmonic oscillator.

The operators Q and P satisfy the commutation relations

$$\left[Q_k, P_{k'}\right] = i\hbar\, \delta_{kk'}. \tag{A.2}$$

We now introduce the dimensionless operators:

$$a_k^\dagger = \frac{1}{\sqrt{2\hbar\omega_k}}(\omega_k Q_k^* - \mathrm{i}P_k)$$

$$a_k = \frac{1}{\sqrt{2\hbar\omega_k}}(\omega_k Q_k^* + \mathrm{i}P_k)$$

and write the Hamiltonian in terms of these operators, substituting them into (A.1):

$$H = \sum_k \left(a_k^\dagger a_k + \frac{1}{2}\right)\hbar\, \omega_k.$$

In this case, the Schrödinger equation can be written as

$$\sum_k \left(a_k^\dagger a_k + \frac{1}{2}\right)\hbar\,\omega_k\,\psi = \sum_k \left(n_k + \frac{1}{2}\right)\hbar\,\omega_k\,\psi, \qquad n_k = 0, 1, 2, \ldots$$

We can now write the following relation:

$$a_k^\dagger a_k\,\psi = n_k\,\psi.$$

$a_k^\dagger a_k$ can be interpreted as the *particle number operator* and n_k its eigenvalue. We could use the symbol n_k for the number operator, but it should not be confused with its eigenvalue. The eigenvalue is a real number, while the operator is just a symbol. An eigenvalue equation for this operator acting on the wave function can be written as

$$a_k^\dagger a_k\,|n_1 \ldots n_k \ldots\rangle = n_k\,|n_1 \ldots n_k \ldots\rangle.$$

Using equation (A.2), we can derive the commutation relation for the $a_k^\dagger$ and a_k operators:

$$[a_k, a_{k'}^\dagger] = a_k a_{k'}^\dagger - a_{k'}^\dagger a_k = \delta_{kk'}. \tag{A.3}$$

The $a_k^\dagger$ and a_k operators have the following effects, respectively:

$$\begin{aligned} a_k^\dagger\,|n_k\rangle &= \sqrt{n_k + 1}\,|n_k + 1\rangle \\ a_k\,|n_k\rangle &= \sqrt{n_k}\,|n_k - 1\rangle \end{aligned}. \tag{A.4}$$

It can also be shown that (for any k)

$$\begin{aligned} &[a, a^\dagger] = aa^\dagger - a^\dagger a = 1 \\ &[aa^\dagger, H] = [a^\dagger a, H] = a^\dagger a\,H - H\,a^\dagger a = 0 \end{aligned}. \tag{A.5}$$

From the last expression, we can see that the Hamiltonian commutes with $a^\dagger a$ and therefore has the same eigenfunctions, i.e.

$$\begin{aligned} H\,\psi_n &= E_n\,\psi_n \\ a^\dagger a\,\psi_n &= n\,\psi_n. \end{aligned} \tag{A.6}$$

Here, ψ_n form the complete set of solutions of the Schrödinger equation of the harmonic oscillator.

From (A.5) and (A.6) we can calculate

$$a^\dagger a(a^\dagger\psi_n) = a^\dagger(aa^\dagger\psi_n) = a^\dagger(n + 1)\psi_n = (n + 1)(a^\dagger\psi_n)$$

concluding that

$$a^\dagger a(a^\dagger\psi_n) = (n + 1)(a^\dagger\psi_n),$$

which implies that the operator $a^\dagger$ applied to ψ_n increases the value n by one. In fact,

$$a^\dagger\psi_n = \sqrt{n + 1}\,\psi_{n+1}.$$

The factor $\sqrt{n+1}$ is obtained by requiring that ψ_{n+1} be normalized and that the factor c relating ψ_n to ψ_{n+1} be real: taking $a^\dagger\psi_n = c\,\psi_{n+1}$ and using $\langle\psi_{n+1}|\psi_{n+1}\rangle = 1$, we get

$$\langle\psi_{n+1}|\psi_{n+1}\rangle = 1 = (1/c^2)\langle a^\dagger\psi_n|a^\dagger\psi_n\rangle = (1/c^2)\langle\psi_n|aa^\dagger|\psi_n\rangle$$

and

$$\langle aa^\dagger\rangle = n + 1.$$

Therefore,

$$\langle\psi_{n+1}|\psi_{n+1}\rangle = (1/c^2)\langle\psi_n|n+1|\psi_n\rangle = \frac{n+1}{c^2}, \quad \langle\psi_n|\psi_n\rangle = \frac{n+1}{c^2} = 1,$$

from which we get

$$c = \sqrt{n+1}.$$

In the same way, we can show that a is an operator, such that

$$a\,\psi_n = \sqrt{n}\,\psi_{n-1}.$$

The operators $a^\dagger$ and a are the called *creation* and *annihilation* operators, respectively, because, when applied to a wave function, they create or annihilate a quantum of energy or a particle (e.g. $\hbar\omega$).

All this implies that we can get the eigenvector $|n_1 n_2 \ldots n_k \ldots\rangle$ by successively applying the creation operator $a_k^\dagger$ to the vacuum state $|00\ldots0\ldots\rangle$:

$$|n_1\, n_2\, \ldots n_k \ldots\rangle = \ldots(a_k^\dagger)^{n_k}\ldots(a_1^\dagger)^{n_1}|0\,0\,\ldots0\ldots\rangle. \tag{A.7}$$

Note that the commutation relation (A.3), from which many properties of the $a^\dagger$ and a operators can be derived, is only valid for bosons (e.g. phonons, photons).

We can also introduce an equivalent representation for fermions. As a consequence of Pauli's exclusion principle, each state can only be occupied by one fermion. The n_k in the eigenvectors can only have values of zero or one. We now have the creation and annihilation operators of fermions, which we denote by $c_k^\dagger$ and c_k, respectively:

$$c_k^\dagger\,|0\rangle = |1\rangle, \quad c_k^\dagger\,|1\rangle = 0, \quad c_k\,|0\rangle = 0, \quad c_k\,|1\rangle = |0\rangle. \tag{A.8}$$

Pauli's exclusion principle also implies that the eigenvectors have to be *antisymmetric*—they have to change sign upon the exchange of two particles. To define this exchange, we arrange the n_k, for example, according to the magnitude of the corresponding eigenvalues. The equation for fermions which is equivalent to equation (A.7) for the construction of an eigenvector from the vacuum, can be written, for example, as

$$|1_1\, 1_2\, 1_3\ldots1_k\ldots\rangle = (-1)^{n+1}\ldots c_k^\dagger\ldots c_3^\dagger\, c_2^\dagger\, c_1^\dagger\,|0_1\, 0_2\, 0_3\,\ldots0_k\ldots\rangle$$

which should be read according to the way particles are successively created in states 1, 2, 3, ...k.... The order of the states is now important, because the exchange of two particles means the exchange of n_k, n_k' in the eigenvector, and therefore, accordingly, to the exchange of c_k, c_k' on the right-hand side. In order for the exchange to imply a change of sign, we require that

$$c_k^\dagger c_{k'}^\dagger = -c_{k'}^\dagger c_k^\dagger \tag{A.9}$$

and equivalently

$$c_k c_{k'} = -c_{k'} c_k, \quad c_k c_{k'}^\dagger = -c_{k'}^\dagger c_k, \quad (k \neq k'). \tag{A.10}$$

For $k = k'$, the relation (A.9) and the first of the equations (A.10) are equally valid, since according to (A.8), the products $c_k^\dagger c_k^\dagger$ and $c_k c_k$ are equal to zero.

From (A.8), we obtain the following for the second equation of (A.10):

$$c_k c_k^\dagger|0\rangle = |0\rangle, \quad c_k^\dagger c_k|0\rangle = |0\rangle, \quad c_k c_k^\dagger|1\rangle = |0\rangle, \quad c_k^\dagger c_k|1\rangle = |1\rangle,$$

and consequently

$$c_k c_k^\dagger + c_k^\dagger c_k = 1 \quad \text{or} \quad [c_k, c_{k'}^\dagger]_+ = \delta_{kk'}. \tag{A.11}$$

We conclude that, in the commutation relations for fermions, the *commutators* are replaced by the *anticommutators* $[c_k, c_{k'}^\dagger]_+$ shown in equation (A.11). The relations (A.8) and the commutation relations lead to equations analogous to (A.4) which describe the effect of the operators:

$$\begin{aligned} c_k^\dagger\,|...n_k...\rangle &= \sqrt{1-n_k}\,(-1)^{\nu_k}\,|...n_k+1...\rangle \\ c_k\,|...n_k...\rangle &= \sqrt{n_k}\,(-1)^{\nu_k}\,|...n_k-1...\rangle \end{aligned},$$

where $\nu_k = \sum_{i<k} n_i$. So, we have a positive sign when there are an even number of states in the eigenvector to the left of n_k and a negative sign when the number is odd.

We now show how to convert the quantum mechanical equations in the $\mathbf{r}$-*representation* to the *occupation number* or *second quantization* representation. To do this, we consider the simplest case of a Hamiltonian H built from a sum of single-particle Hamiltonians $h(\mathbf{r}_i)$: $H = \sum_i h(\mathbf{r}_i)$.

For bosons, the wave function, which is invariant against an exchange of particles, is the sum of all particle permutations:

$$\Phi = \frac{1}{\sqrt{N!n_1!n_2!\ldots}} \sum_P P\, \phi_\alpha(\mathbf{r}_1)\ldots\phi_\alpha(\mathbf{r}_{n1})\ldots\phi_\omega(\mathbf{r}_N),$$

where the sum extends to all permutations of the indices α, ...ω, and where n_1 factors have the index α, n_2 factors have the index β, etc. The Hamiltonian can then be written as

$$\begin{aligned} H &= \textstyle\sum_i h(\mathbf{r}_i) \\ H &= \textstyle\sum_{\lambda\lambda'} \langle\lambda'|h|\lambda\rangle\, a_k^\dagger a_k \end{aligned},$$

where

$$\langle \lambda' | h | \lambda \rangle = \int \phi^*_{\lambda'}(\mathbf{r})\, h\, \phi_\lambda(\mathbf{r}) \mathrm{d}\tau.$$

For fermions, the wave function changes sign every time two particles are exchanged. Therefore, the wave function is the sum of all permutations of pairs of particles with alternate plus and minus signs:

$$\Psi = \frac{1}{\sqrt{N!}} \sum_P (-1)^P \, \phi_{k_1}(\mathbf{r}_1)\phi_{k_2}(\mathbf{r}_2)\ldots\phi_{k_N}(\mathbf{r}_N),$$

where P is the number of permutations. This can also be written as a Slater determinant:

$$\Psi(\mathbf{r}_1, \mathbf{r}_2, \ldots, \mathbf{r}_N) = \frac{1}{\sqrt{N!}} \begin{vmatrix} \phi_{k_1}(\mathbf{r}_1) & \phi_{k_1}(\mathbf{r}_2) & \ldots & \phi_{k_1}(\mathbf{r}_N) \\ \phi_{k_2}(\mathbf{r}_1) & \phi_{k_2}(\mathbf{r}_2) & \ldots & \phi_{k_2}(\mathbf{r}_N) \\ \ldots & \ldots & & \ldots\ldots \\ \phi_{k_N}(\mathbf{r}_1) & \phi_{k_N}(\mathbf{r}_2) & \ldots & \phi_{k_N}(\mathbf{r}_N) \end{vmatrix}, \tag{A.12}$$

which, by convention, can be written in the second quantization as:

$$\Psi(\mathbf{r}_1, \mathbf{r}_2, \ldots, \mathbf{r}_N) = c^\dagger_{k_N} c^\dagger_{k_{N-1}} \ldots c^\dagger_{k_2} c^\dagger_{k_1} |0\rangle = \prod_{k_i} c^\dagger_{k_i} |0\rangle. \tag{A.13}$$

This representation is interpreted as the successive creation, from the vacuum, of fermions of momentum $\mathbf{k}_i$. The antisymmetry of the wave functions is considered implicit, so that (A.13) represents the Slater determinant (A.12).

The conjugate wave function of Ψ is written as

$$\Psi^*(\mathbf{r}_1, \mathbf{r}_2, \ldots, \mathbf{r}_N) = \langle 0 | c_{k_1} c_{k_2} \ldots c_{k_{N-1}} c_{k_N},$$

and the Hamiltonian as:

$$H = \sum_i h(\mathbf{r}_i)$$
$$H = \sum_k \epsilon_k \, c^\dagger_k c_k,$$

where ϵ_k are the eigenvalues of the particles' energy.

In turn, the matrix elements involving two electronic states can be written as

$$V(\mathbf{r}) = \sum_{k,\, k'} \langle \mathbf{k}' | V | \mathbf{k} \rangle \, c^\dagger_{k'} c_k, \quad \langle \mathbf{k}' | V | \mathbf{k} \rangle = \int \phi^*_{k'}(\mathbf{r}) \, V \, \phi_k(\mathbf{r}) \, d\tau,$$

and the electron–electron interaction can be written as

$$V(\mathbf{r}_1, \mathbf{r}_2) = \frac{1}{2} \sum_{k_1, k_2, k_3, k_4} \langle k_4, k_3 \mid V \mid k_2, k_1 \rangle c^\dagger_{k_4} c^\dagger_{k_3} c_{k_2} c_{k_1},$$

where

$$\langle k_4, k_3 \mid V \mid k_2, k_1\rangle = \int d\tau_1 d\tau_2 \phi^*_{k_4}(\mathbf{r}_1)\phi^*_{k_3}(\mathbf{r}_2)\, V(\mathbf{r}_1 - \mathbf{r}_2)\, \phi_{k_2}(\mathbf{r}_2)\phi_{k_1}(\mathbf{r}_1).$$

A final note: The occupation number states $|n_\alpha\rangle$ are also known as Fock states. All the Fock states form a complete basis of the many-body Hilbert space, or Fock space.

Further reading

Giamarchi T 2004 *Quantum Physics in One Dimension* (Oxford: Clarendon) appendix A

IOP Publishing

The Physics of Organic Electronics
From molecules to crystals and polymers
Luís Alcácer

Appendix B

Phase transitions and scaling laws

B.1 Phase transitions

In a phase transition $\alpha \rightarrow \beta$, $\mu_\beta - \mu_\alpha < 0$, where μ_α and μ_β are the chemical potentials (Gibbs energy per mole) of phases α and β. According to Ehrenfest the transitions are labelled by the lowest order of the derivative of μ that is discontinuous for the transition. The first-order transitions have a discontinuity in the first derivative of μ with respect to another variable, such as the temperature. The second-order transitions are continuous with respect to the first derivative, but are discontinuous with respect to the second derivative.

First-order transitions (e.g. melting) occur with abrupt symmetry breaking: the phase of higher symmetry corresponds to $T > T_c$, where T_c is the critical or transition temperature. The specific heat C_p is infinite at T_c; the entropy and the enthalpy are discontinuous, and there is a non-zero enthalpy of transition. In the second-order transitions (e.g. superconductivity), the symmetry changes continuously; $d\mu/dT$ is continuous at T_c. C_p is discontinuous; entropy and enthalpy are continuous. These transitions are characterized by a diverging susceptibility, an infinite correlation length, and an exponential decay of the correlations near a critical point. The continuous change of symmetry is characterized by an *order parameter* ϕ, which is a measure of the degree of order in the system—it changes from zero in the high-symmetry phase (e.g. the normal state) to a saturation value for the ordered phase (e.g. the superconducting state). ϕ is a continuous function of the temperature (see figure B.1).

A remarkable feature of continuous phase transitions is self-similarity, or scale invariance. At the critical points, all systems show similar features on all scales, a fractal structure which emerges from a hidden symmetry—these systems are statistically invariant under a change in length scale. The rescaling process is a renormalization group transformation.

doi:10.1088/978-0-7503-3347-4ch11

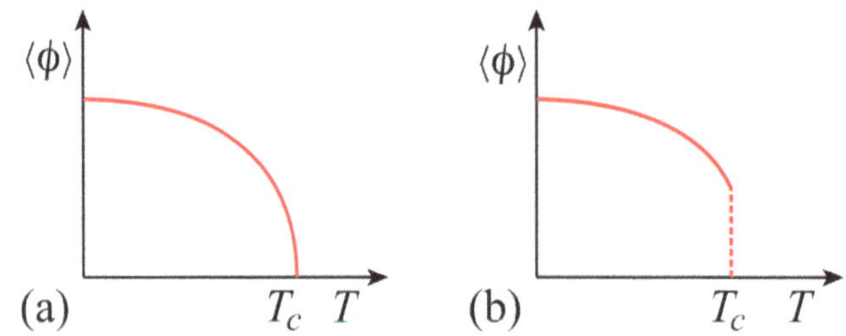

Figure B.1. Order parameter vs transition temperature in second- (a), and first-order (b) phase transitions.

Table B.1. Critical exponents of some thermodynamic properties.

Exponent	α	β	γ	ν	η
Property	Specific heat	Order parameter	Susceptibility	Corr. length	Corr. function
Definition	$C \sim \lvert t\rvert^{-\alpha}$	$\langle\phi\rangle \sim \lvert t\rvert^{\beta}$	$\chi \sim \lvert t\rvert^{-\gamma}$	$\xi \sim \lvert t\rvert^{-\nu}$	$G(q) \sim q^{-2+\eta}$
Mean field	0	0.5	1	0.5	0
Ising	0.11	0.32	1.24	0.63	0.03–0.06
Experimental	0.11	0.32	1.24–1.82	0.6–1.1	0.03–0.06
Ord. mag. (3D)	~0	~1/3	~4/3	~2/3	~0

B.2 Critical exponents

For continuous phase transitions in the vicinity of a critical point, the correlations decay exponentially with the *critical exponents*. An example is the exponent which describes the divergency of the magnetic susceptibility near the ferromagnetic phase transition. The susceptibility diverges as T approaches T_c, following the power law

$$\chi \propto |T - T_c|^{-\gamma},$$

where γ is typically between 1.3 and 1.4.

To study the critical exponents, we define a reduced temperature t:

$$t = \frac{T - T_c}{T_c},$$

which measures the deviation from the critical temperature and a reduced conjugate field h (e.g. a dimensionless magnetic field). Similar behaviours are observed for the correlation length and for other properties, each of which has its characteristic exponent (see table B.1).

The critical exponent λ of a given function is defined as

$$\lambda = \lim_{t \to 0} \frac{\ln|f(t)|}{\ln|t|},$$

which implies

$$f(t)_{t \approx 0} \sim |t|^{\lambda}.$$

This means that near the critical temperature T_c, the function $f(t)$ is dominated by t^{λ}.

A method for calculating the critical exponents was introduced by Kenneth Wilson in 1975 [1] based on a new approach called the 'renormalization group', a procedure which can deal with the problems of renormalization of many lengths and energy scales. This method also establishes that the critical exponents depend on the spatial dimensions of the system, the order parameter, the symmetry and order of magnitude of the interactions, but not on the details of the form and intensity of the interactions. There are *universality classes*, in which all transitions within the class have the same critical exponents.

Another important feature of the second-order phase transitions is that not all critical exponents are independent. For example, γ is always of the order of 2ν, and $\alpha + 2\beta + \gamma$ is of the order of two. These relations are a consequence of the homogeneity of the scaling properties of the correlation and thermodynamic functions near T_c, which can be derived from the renormalization group.

B.3 Scaling laws

If we consider a cube with edge L and scale it by a scaling factor of λ, its surface area $A = 6L^2$ becomes $A' = 6\,\lambda^2 L^2 = \lambda^2 A$ and its volume $V = L^3$ becomes $V' = \lambda^3 L^3 = \lambda^3 V$. In general, we can consider how a given 'property' $M(r)$ of a system changes when its characteristic size r in the embedding space is rescaled by a factor of λ, i.e. $r \to \lambda r$. The scaling law is

$$M(\lambda r) = \lambda^D M(r).$$

If r is a length, then D is the dimensionality of the Euclidean space.

In the cases of the area or the volume, the exponent is an integer. But in Nature there are many situations in which the dimensionality of the embedding space is a fraction. The concept of fractional or 'fractal' dimensions applies to self-similar structures. The term 'fractal' was coined by Benoit Mandelbrot, who described and showed many beautiful fractal structures in his book 'The Fractal Geometry of Nature' [2].

Near the critical points, some thermodynamic functions, such as the free energy and the susceptibility, are homogeneous functions of the temperature and the conjugate field. We can then use the definition of homogeneous functions

$$f(\alpha x, \alpha y) = \alpha^k f(x, y)$$

to determine the scaling relations. The correlation length, for example, is the only relevant length near the critical point and can, therefore, be used as a scale for all relevant distances, namely to determine how the thermodynamic functions vary with temperature.

Scaling laws are pervasive to many topics in the physics of organic conductors and organic electronic devices, namely in phase transitions, conductivity in disordered organic semiconductors, surface growth, surfaces and interfaces, and polymer physics.

Glossary

Correlation function The correlation function between two variables measures how correlated the two variables are. The correlation function between two sets of random variables, (e.g. the density at $\mathbf{x}_1$, $n(\mathbf{x}_1)$, and the density at $\mathbf{x}_2$, $n(\mathbf{x}_2)$) is the average of the product of these two variables:

$$C_{nn}(\mathbf{x}_1, \mathbf{x}_2) = \langle n(\mathbf{x}_1)\, n(\mathbf{x}_2) \rangle.$$

Correlation length The characteristic length, ξ, which diverges as the critical temperature, T_c, of a second-order phase transition is approached: $\xi \sim |T - T_c|^{-\nu}$.

Critical exponents Exponents that characterize the singularities of variables at the critical points of second-order transitions, e.g. $C_p \sim |T - T_c|^{-\alpha}$, $\phi \sim |T - T_c|^{-\beta}$.

Critical phenomena The emergence of order at continuous phase transitions characterized by the formation of short-lived droplets of nascent order that grow as the system is tuned at the critical point. At the critical point, droplets of all sizes of one phase are interspersed among droplets of the other phase, across the material.

Fractal A fractal is a pattern that repeats itself at different scales. The term 'fractal' (from the Latin 'fractus') was coined by the mathematician Benoit Mandelbrot in 1975 to describe the theoretical *fractional dimensions* of geometric structures in Nature.

Heisenberg model Model for a spin system $\mathbf{s}_i$ pointing in any direction, leading to a magnetization $\langle \mathbf{m} \rangle$ that is not fixed relative to the lattice. This model undergoes a second-order phase transition between an ordered and a disordered phase. It is disordered above a critical temperature, and it exhibits ferromagnetic order below the critical temperature.

Ising model A lattice of N sites with a spin σ_i on each site. Each spin can take one of two possible values: +1 for spin up and −1 for spin down. There are a total of 2^N possible configurations of the system. A configuration is specified by the orientations of the spins on all N sites: $\{\sigma_i\}$, where σ_i is the spin on the ith lattice site. The order parameter is the magnetization $\langle m_z \rangle$. In one dimension this model has an exact solution, showing that a second-order critical point can be considered at $T = 0$, leading to a divergent Curie susceptibility, $\chi \sim T^{-1}$, and a divergent correlation length ξ.

Magnetization The magnetization (magnetic moment per unit volume in the presence of an applied magnetic field) of an ideal gas of particles, each with spin $\mathbf{s}_i$, and a magnetic moment $\mu_s \mathbf{s}_i$, is $\langle \mathbf{m}(\mathbf{x}) \rangle = \langle \sum_i \mu_s \mathbf{s}_i \delta(\mathbf{x} - \mathbf{x}_i) \rangle$.

Magnetization is an operator like the density operator. (Note: $\mathbf{s} = S_z$ is the spin operator with quantum numbers $m_s = \pm 1/2$. The magnetic moment per particle is $\mu_s = -g m_s \mu_B$, $g = g$ factor, $\mu_B =$ Bohr magneton). At the critical temperature T_c, there is a phase transition from the disordered paramagnetic state to the *ordered* ferromagnetic state. The magnetization $\langle \mathbf{m} \rangle$ is the *order parameter* (figure B.1(a)). Below T_c, the magnetization grows continuously from zero up to a saturation value at $T = 0$.

Number density operator Number of particles per unit volume $n(\mathbf{x}) \equiv \sum_i \delta(\mathbf{x} - \mathbf{x}_i)$, where $\mathbf{x}_i$ is the position of particle i. In homogeneous isotropic media, $\langle n(\mathbf{x}) \rangle = n = N/V$ ($N =$ number of particles, $V =$ volume). The sum over the δ function has units of density. $\langle n(\mathbf{x}) \rangle$ has rotational and translational invariance (symmetry). In crystals, $\langle n(\mathbf{x}) \rangle$ is a periodic function of $\mathbf{x}$ and rotational and translational invariance are broken.

Order parameter An order parameter is a label (or a field) for a broken symmetry state of a system which describes its approximate microscopic configuration. It characterizes the continuous change of symmetry in a second-order phase transition. In general, the order parameter is the average, $\langle \phi \rangle$, of an operator ϕ. The order parameter is zero in the disordered phase and non-zero in the ordered phase. It should be chosen to have values that reflect the symmetry of the Hamiltonian, i.e. it should transform under an irreducible representation of the symmetry group of the Hamiltonian. One of the easiest order parameters to understand is that of a simple model of a magnet. By becoming a magnet, a material breaks rotational symmetry. The order parameter $\langle \mathbf{m} \rangle$ labels whichever of the various broken symmetry directions the material has chosen. In superconductivity, the order parameter is the wave function $\psi = |\psi| \exp(i\phi)$, or the gap—the real part of the order parameter—which opens gradually as T is reduced below T_c.

Phase transition A transformation occurring at the unstable interface between two stable phases of matter due to the collective motion of electrons and atoms.

Quantum criticality A situation in which a continuous phase transition occurs at temperatures low enough (in theory, $T = 0$) so that quantum mechanics plays a crucial role. Typically, a 'V-shaped' phase of quantum critical matter spreads out from the quantum critical point at $T = 0$ to higher temperatures.

Quantum phase transition Second-order phase transition at the absolute zero of temperature ($T = 0$) as a result of competing ground states. Such transitions are driven by the quantum fluctuations required by the Heisenberg uncertainty principle.

Renormalization group 'The idea is that whenever one has a problem with many length scales, or many energy scales, etc, one breaks up the calculation into separate stages, with one stage for each length or energy scale. This break-up is to be defined so that each stage involves only a few lengths or energy scales' [1].

Scaling laws Laws that describe the scale invariance found in many natural phenomena. For example, near the critical point, the correlation length can be used as a scale for all distances. If we know how the correlation length varies with the temperature, we can find out how the thermodynamic functions and correlations vary with the temperature.

Symmetry breaking Symmetry means 'invariance under a specified group of transformations, such as rotations, translations, or scaling'. For example, the different phases of matter differ in their symmetry: in the transition from liquid water to solid water, there is *symmetry breaking* from a higher symmetry in liquid water to a lower symmetry in ice (contrary to our idea that a crystal of ice looks more symmetric). While liquid water has complete rotational and translational symmetry, ice has a discrete translational and rotational symmetry due to the regular arrangement of atoms in the crystal.

Universality When two different systems have the same singular properties (e.g. power-law singularities) at their critical points, we say they are in the same *universality* class.

Further reading

Chaikin P M and Lubensky T C 1995 *Principles of Condensed Matter Physics* (Cambridge: Cambridge University Press)

Sethna J P 2017 *Statistical Mechanics: Entropy, Order Parameters, and Complexity* 2nd edn (Oxford: Clarendon)

References

[1] Wilson K G 1975 *Adv. Math* **16** 170–86

[2] Mandelbrot B B 1983 *The Fractal Geometry of Nature* (New York: Freeman)

Appendix C

Atomic units

In quantum chemical calculations it is convenient to use *atomic units* (au or a.u.), which are based on the following four fundamental constants taken as unity by definition:

Electron mass	$m_e = 1$
Elementary charge	$e = 1$
Reduced Planck's constant	$\hbar = 1$
Coulomb's constant	$k_0 = 1/(4\pi\varepsilon_0) = 1$

Maxwell's equations

Maxwell's macroscopic equations, also known as Maxwell's equations in matter, are (in SI units):

$$\nabla \cdot \mathbf{D} = \rho \qquad \text{Gauss's law}$$
$$\nabla \cdot \mathbf{B} = 0 \qquad \text{Gauss's magnetism law}$$
$$\nabla \times \mathbf{E} = -\frac{\partial \mathbf{B}}{\partial t} \qquad \text{Faraday's law}$$
$$\nabla \times \mathbf{H} = \mathbf{J} + \frac{\partial \mathbf{D}}{\partial t} \qquad \text{Ampere's law.}$$

Other fundamental relations are:

$$\mathbf{D} = \varepsilon_0\mathbf{E} + \mathbf{P} = \varepsilon\mathbf{E} \qquad \mathbf{B} = \mu_0(\mathbf{H} + \mathbf{M}) = \mu\mathbf{H}$$

E = Electric field, ρ = Electric charge density, **D** = Electric flux density, **B** = Magnetic flux density, **H**= Magnetic field, **P**= Polarization, **M** = Magnetization, ε = Permittivity, ε_0 = Vacuum permittivity, μ = Magnetic permeability, μ_0 = Vacuum magnetic permeability.

NOTE 1: The term 'magnetic field' is used for two distinct fields, **B** and **H**. In SI units, **H**, magnetic field strength, is measured in amperes per meter ($\mathrm{A\,m^{-1}}$). **B**, magnetic induction or magnetic flux density, is measured in tesla; $\mathbf{B} = \mu\mathbf{H} = \mu_0\mu_r\mathbf{H}$.

NOTE 2: Historically, Maxwell's equations predicted that electromagnetic radiation travelled at a speed given by $\sqrt{1/\varepsilon_0\mu_0}$. This value is equal to the speed of light, suggesting that light is the electromagnetic radiation predicted by Maxwell's equations.

Appendix D

Problem hints and solutions

2.1 (b) $n = 4.56 \times 10^{25}$ m^{-3}; (c) $m^{*}=0.819\ m_0$.

2.2 $\mathbf{v} \times \mathbf{B} = B\begin{pmatrix} v_y \\ -v_x \end{pmatrix} = B\begin{pmatrix} 0 & 1 \\ -1 & 0 \end{pmatrix}\begin{pmatrix} v_x \\ v_y \end{pmatrix}$, and the corresponding diagonal matrix preceding the column vector $\mathbf{E} = (E_x, E_y)$.

3.1 Hint: use Gauss' theorem (3.2), which defines the flux Φ (number of field lines) leaving a closed surface S_2 (of any shape). We can always imagine a spherical surface S_1 containing the charge q and write $\oint_{S_2} \mathbf{E} \cdot \mathrm{d}\mathbf{A} = \oint_{S_1} \mathbf{E} \cdot \mathrm{d}\mathbf{A} = q/\varepsilon$. Now consider many charges in a unit volume defining a charge density $\sum_i q_i/\varepsilon = (1/\varepsilon)\int_V \rho \mathrm{d}\mathbf{r} = \oint_{S_1} \mathbf{E} \cdot \mathrm{d}\mathbf{A}$ and use (3.2) to get $\nabla \cdot \mathbf{E} = \rho/\varepsilon$.

3.2 (a) V = 0.144 kcal mol^{-1}; (b) 0.221 kcal mol^{-1}; (c) at $\sigma = 0.34$ nm; (d) 188 pm.

4.1 (a) Only the planes whose Miller indices' sum is even originate diffraction peaks (equation 4.14).

6.1 (a) $S = 35.74$ μV K^{-1}. Note that $\rho = 3/2$ and $W = 4t$.
(b) $S = 35.74$ μV K^{-1}.
(c) $E_F = 0.103$ eV.

7.1 (a) 0.952 6 nm^3.
(b) 1.16.

7.2 (a) Using (7.8): $E_1 = 6.18$ meV; $E_2 = 1.55$ meV; $E_3 = 0.69$ meV.
(b) $a_{\mathrm{exc}} = 11.6$ nm.

7.3 (a) Using equation (7.11), $R_0 = 7.4$ nm.
(b) $R_{\mathrm{DA}} = R_0 = 7.4$ nm.

8.3 (a) 10%.
(b) 46.3 m^2.

8.4 (a) $C_v = 25$ F cm^{-3}; $C_i = 500$ μF cm^{-2}.

9.2 (a) $N = 9.67 \times 10^{14}$ s^{-1}.
(b) $r = 157$ pm.

doi:10.1088/978-0-7503-3347-4ch13

www.ingramcontent.com/pod-product-compliance
Ingram Content Group UK Ltd.
Pitfield, Milton Keynes, MK11 3LW, UK
UKHW051852100926
8556IPUK00005B/594

9 780750 333450